W9-AIB-543

THE WAR IN IRAQ

THE WAR IN IRAQ

A Photo History

HOWARD COUNTY LIBRARY
BIG SPRING, TEXAS

ReganBooks
An Imprint of HarperCollins*Publishers*

THE WAR IN IRAQ. Copyright © 2003 by ReganBooks. All rights reserved.
Printed in the United States of America. No part of this book may be used or reproduced in any manner whatsoever without written permission except in the case of brief quotations embodied in critical articles and reviews.
For information address HarperCollins Publishers Inc., 10 East 53rd Street, New York, NY 10022

HarperCollins books may be purchased for educational, business, or sales promotional use. For information please write: Special Markets Department, HarperCollins Publishers Inc., 10 East 53rd Street, New York, NY 10022.

FIRST EDITION

Designer: Gabriel Kuo/BRM
Photo Editor: Larry Schwartz

Jacket/cover design by Lauren Prigozen and Susan Carroll
Front jacket/cover photograph by Oleg Popov/Reuters
Hardcover back jacket photograph by Kia Pfaffenbach/Reuters
Paperback back cover photograph by Newsha Tavakolian/Polaris
Hardcover back flap photograph by Kuni Takahashi/*Boston Herald*/ReflexNews

Printed on acid-free paper

Library of Congress Cataloging-in-Publication Data has been applied for.

ISBN 0-06-058286-3 (hardcover)
ISBN 0-06-058437-8 (paperback)

03 04 05 06 07 ❖/QW 10 9 8 7 6 5 4 3 2 1 (hardcover)
03 04 05 06 07 ❖/QW 10 9 8 7 6 5 4 3 2 1 (paperback)

FOR THOSE WHO LIVED AND DIED FOR A FREE IRAQ

INTRODUCTION

The real war will never get in the books, Walt Whitman wrote late in life, reflecting on the great war of liberation that had shaped the American age he lived in. In one sense, of course, he was right: uncounted stories are left behind in war, lost to history even as we remember the fallen heroes who lived them. And yet, ever since the Civil War of which Whitman wrote, the truths and triumphs of war have been captured and preserved—through the lens of the photojournalist.

And never has the drama of war been captured more thoroughly, or unforgettably, than in the work of the hundreds of photographers who covered Operation Iraqi Freedom. From the first strikes on Baghdad through the end of combat announced on May 1, 2003, this new war of liberation lasted only forty-one days. Yet the emotions woven through it are ageless, as the photographs in *The War in Iraq* so eloquently attest. Here, in more than 250 stunning images, are the stories of this modern war played out against an ancient desert backdrop: the heartless Iraqi dictator, and the resolute American president who ended his reign of terror; the eerie crimson isolation of the early sandstorms, and the surreal spectacle of Saddam's ruined palaces; the convoys of American troops prepared for fierce resistance, and the Iraqis who greeted them—some with hostile fire, but many more with white flags and faces brightened with relief.

The faces of this war, like those of any war, may be what we remember most. And in these photographs they look toward us from half a world away, with every expression imaginable. A young American

girl bids farewell to her mother as she goes off to battle; a young boy in Basra revels amid a windfall of pure sugar after years of deprivation. Joyous Iraqi men welcome the troops with homemade American flags, and topple images of Saddam with anger and euphoria. An American soldier at rest is watched with curiosity by two elderly women in a nearby doorway, their traded whispers almost audible. Flush with victory, coalition troops romp unabashed through the dictator's abandoned living quarters. A beaming British servicewoman takes a moment to greet a local schoolgirl in Umm Qasr. A twelve-year-old boy, orphaned by the war and badly burned, confronts our gaze.

As the first air strikes landed in Baghdad, the words "shock and awe" were used to characterize their impact. Now, in the earliest weeks of Iraqi freedom, those simple words have taken on fresh meanings. The American public may have been shocked at how quickly the Iraqi defenses crumbled. The Iraqi people now face the awesome opportunity and challenge of remaking their nation. As Whitman suggested, it may be impossible to capture in words every emotional nuance of an event as tumultuous as a war. But the photographers who risked their lives to chronicle Operation Iraqi Freedom have done history a profound service: on digital files and videotape and 35mm film, they have compiled a crystalline visual record of this extraordinary moment in our history.

They have got it in the books.

The Editors

THE ROAD TO WAR

الله
أكبر

UN
UN

ADDAM HUSSEIN
UN

نعم نعم
للقائد صدام حسين

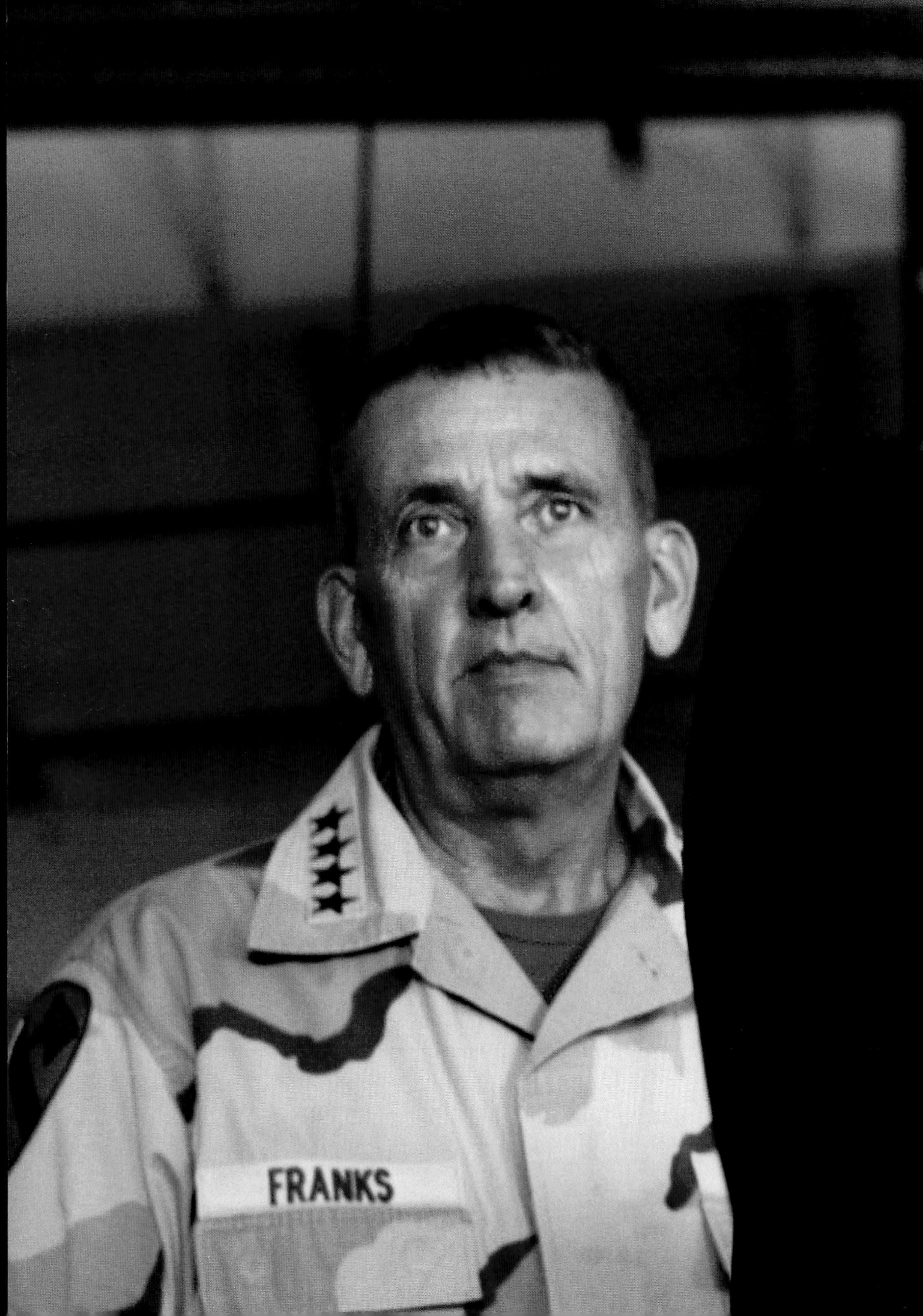
FRANKS

6P

AIRBORNE

محمد رسول الله
سورة الأنفال
بسم الله الرحمن الرحيم

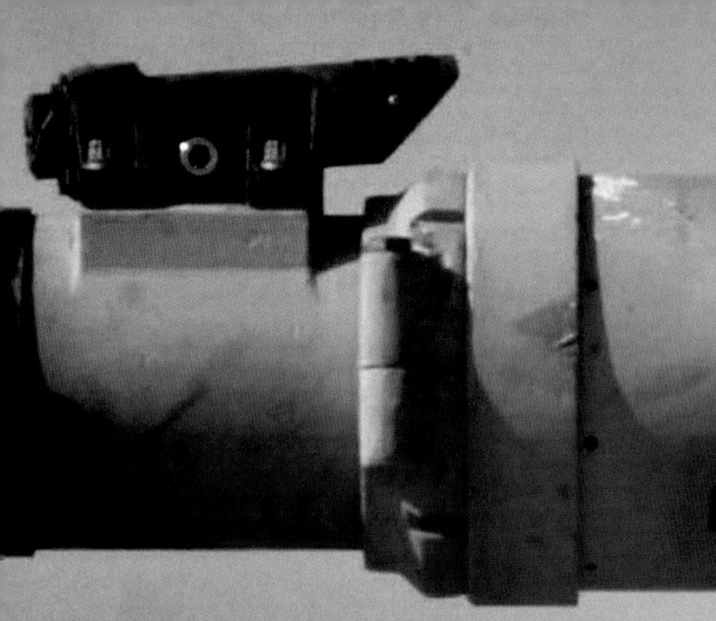

A 9049 ARM BN

WAY TO BAGHDA

VFA 25
FIST OF THE FLEET
400
NK

303
VFA-113
FIST OF THE FLEET
VFA-25
VFA-25
FIST
USS ABRAHAM LINCOLN
VFA-25
LCDR BRIAN KILTY
"PUNCHY"

US

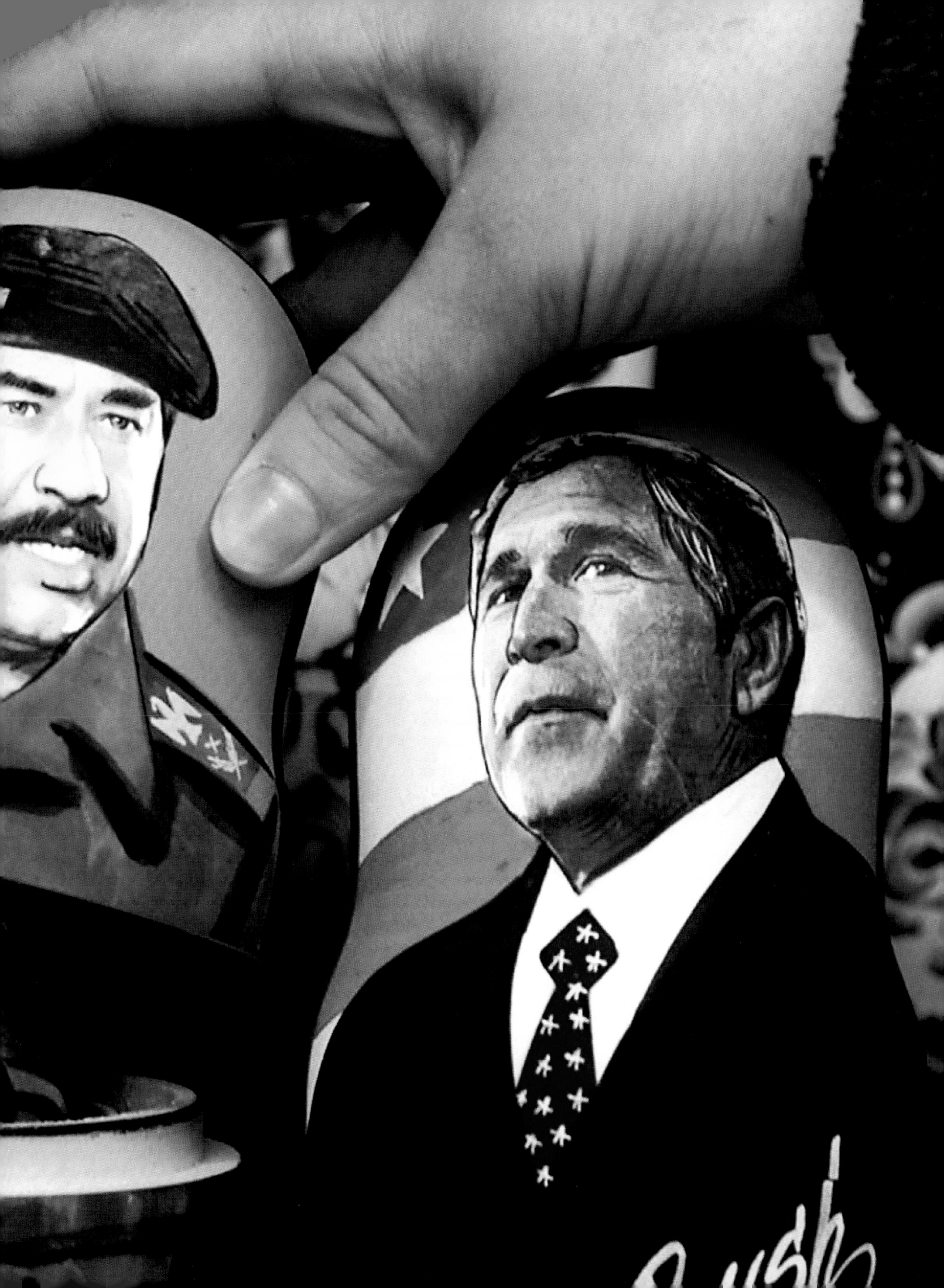

THANK
CLINTON FOR THE
MESS
THANK GOD
FOR

PEACE
THROUGH
STRENGTH

NEW YORK
MADRID
PA
NON
A LA GUERRE IMPERIALISTE
SOLIDARITE
POUR LA PA
AGAINST
de la guerre du Golfe
IS ÇA !

IS
LONDON
BERLIN
AMERICANS
AGAINST
IRAQ
Ausreiseverbot für die Bundeswehr

WAR
PESTILENCE
BLOOD FOR OIL

ORIENT
QUARTZ
JAPAN-5D
58518024-L18

LIVE
BREAKING NEWS
CNN
BUSH: "EARLY STAGES OF
MILITARY OPERATIONS" BEGIN
صدام حسين متحدثا لشعبه
بعد ساعات من الهجوم على بغداد
YEB 0.61
تعتبر ذلك مخالفا لميثاق الأمم المتحدة

WAR

NB

SKELLHAM

UNITED STATES
BROOKS
U.S. ARMY
TIDEMAND

UNITED STATES
CENTRAL COMMAND
FRANKS
U.S. ARMY
AUSTRALIA
UNITED STATES
CENTRAL COMMAND

جيش العراقي الباسل»

٦ كانون الثاني ذكرى تأسيـ

تسقط امريكا

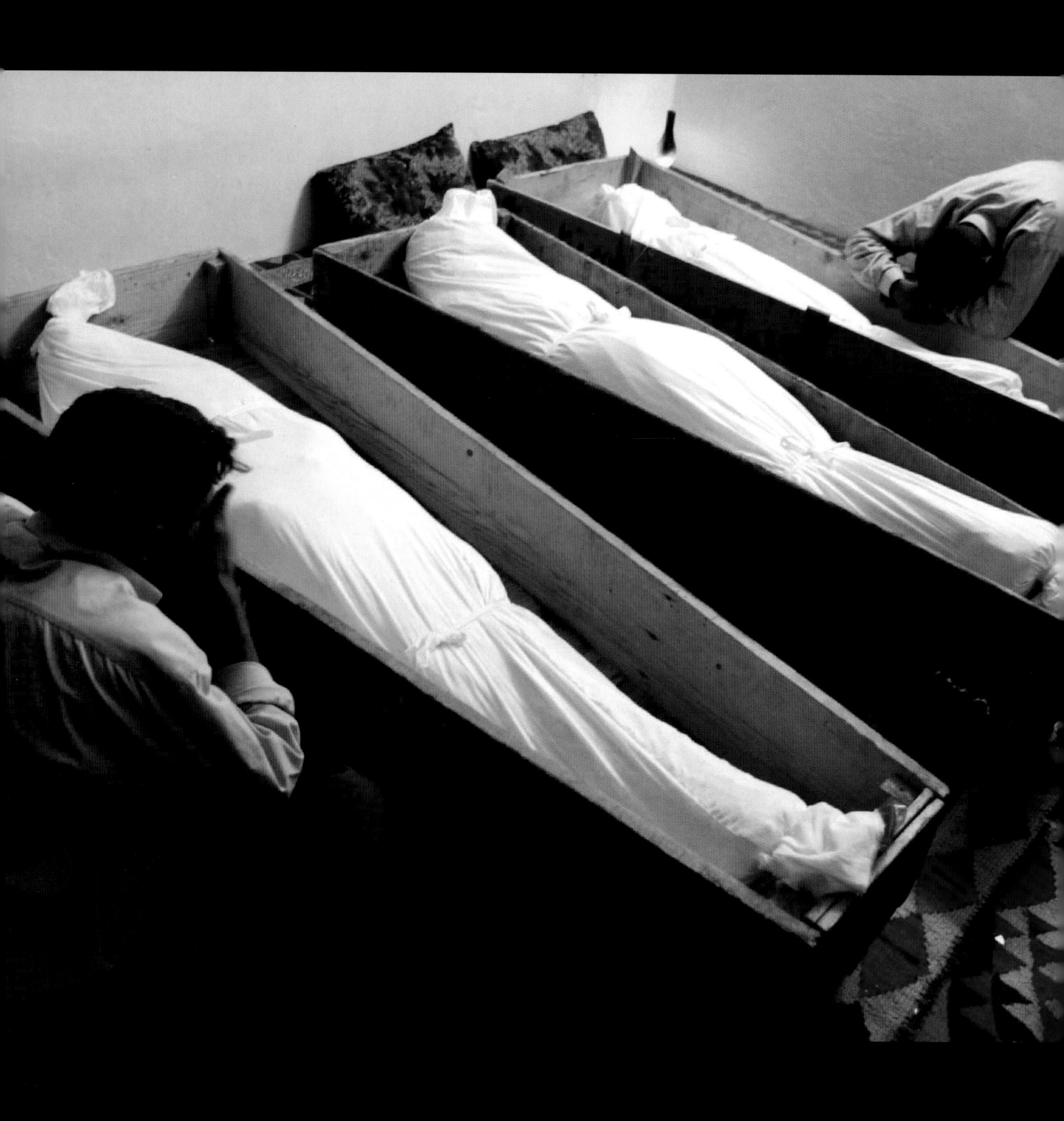

80

Hatta

20

Best

WELCOME
TO BUSH
INTERNATIONAL
AIRPORT
BAR 65,313

XXX
BELGIO

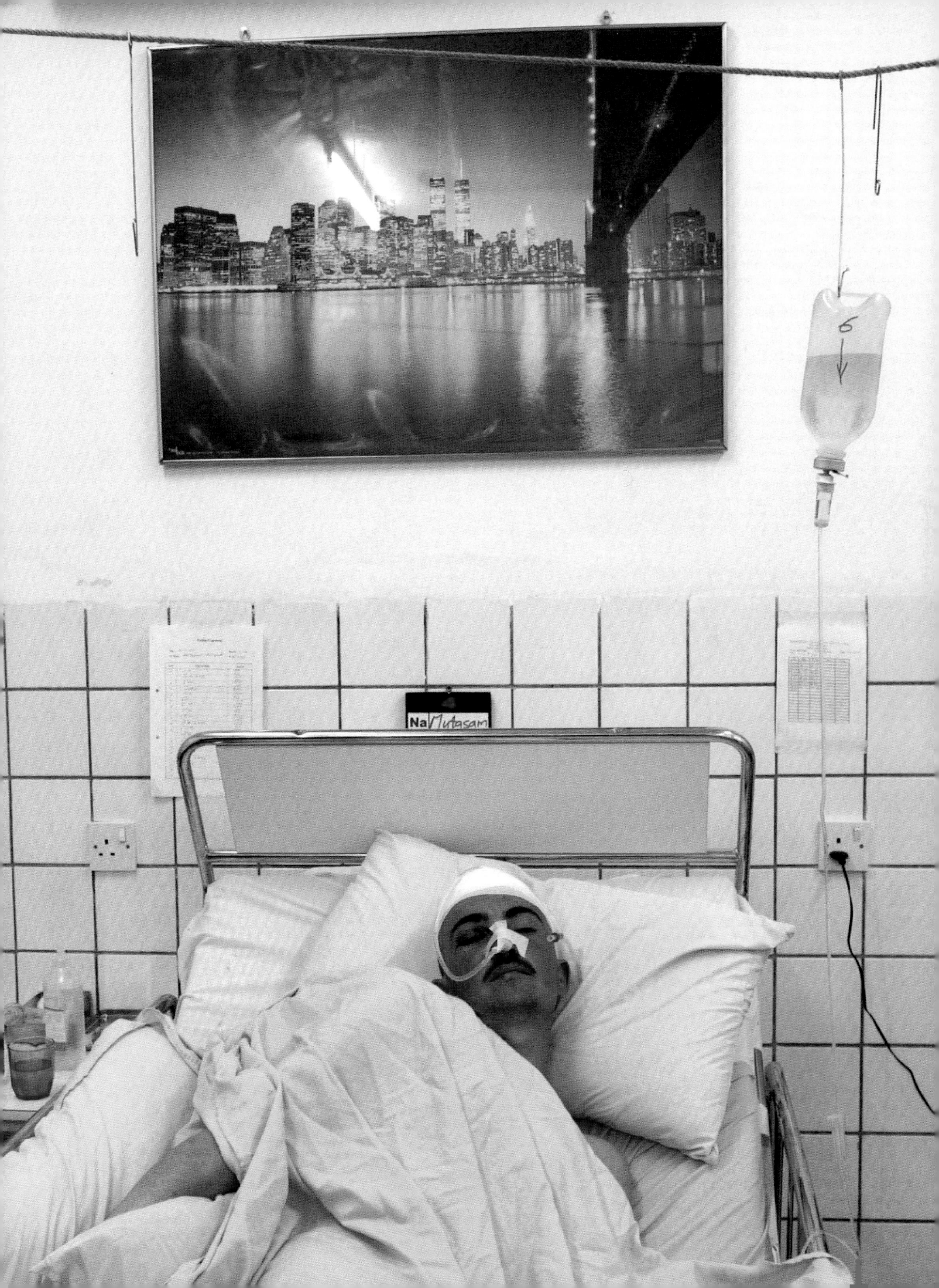
Na Mutasam
6

BUNKER

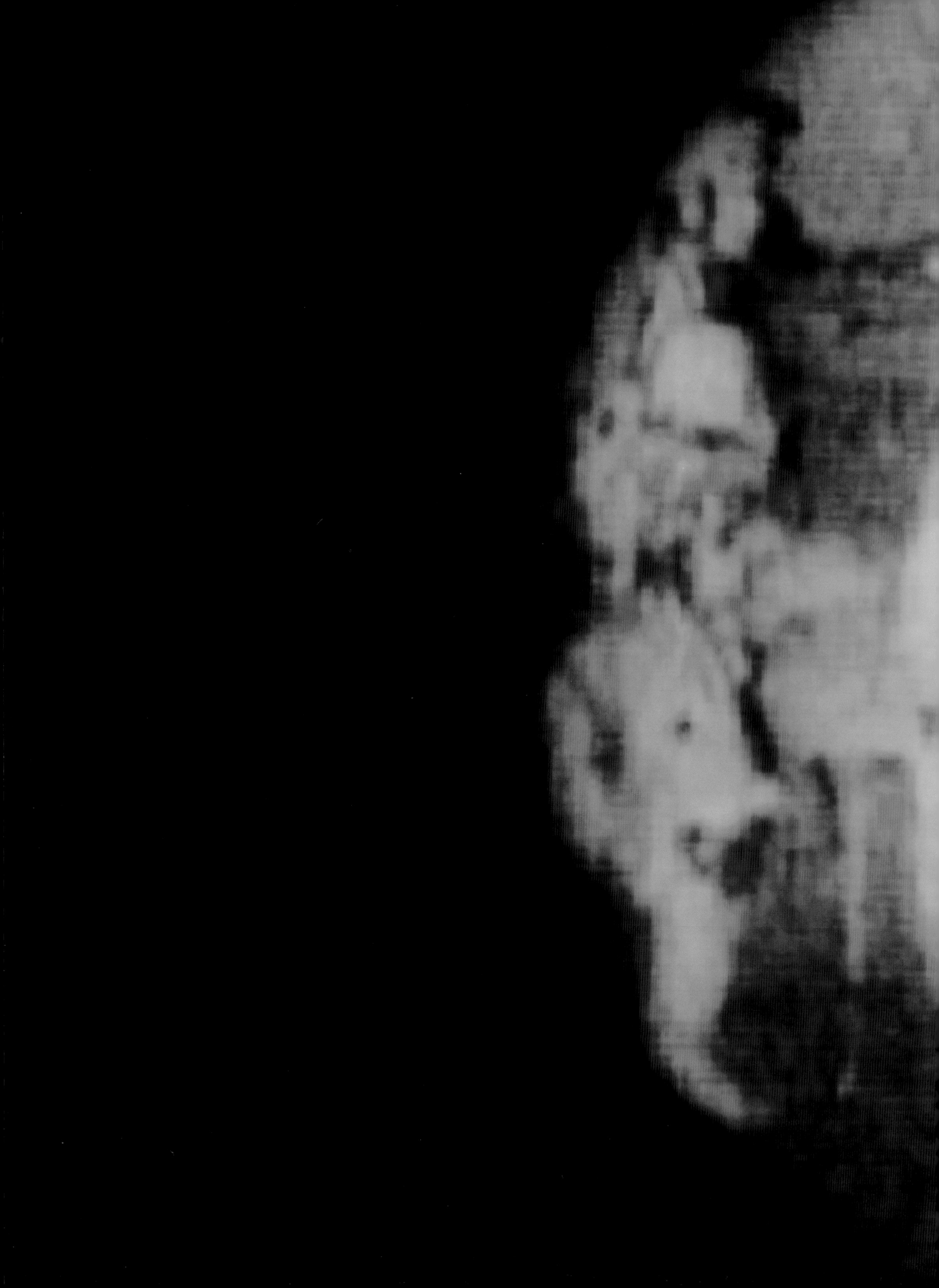

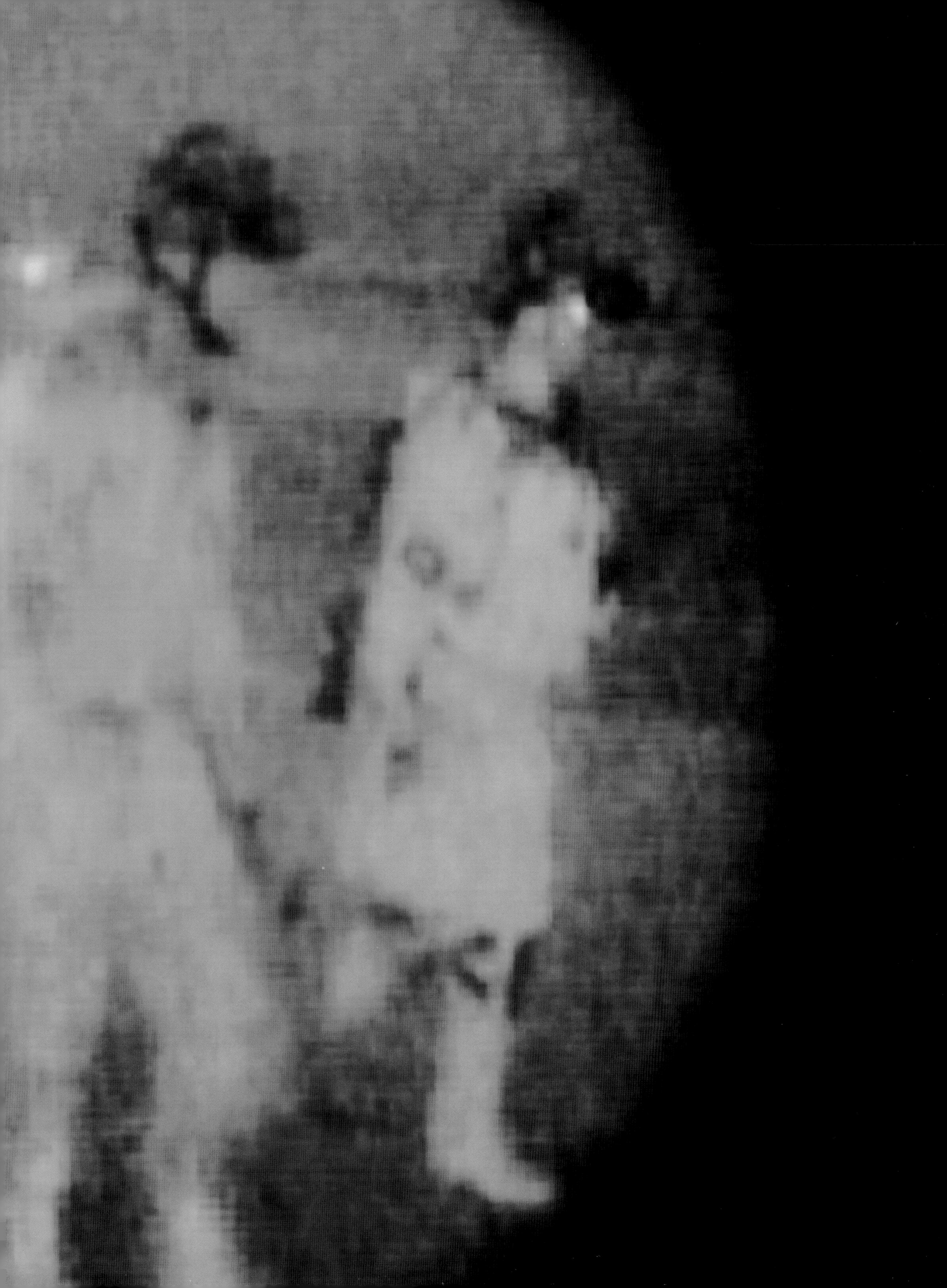

LYNCH
U.S. ARMY

RESCUE
DANGER
EJECTION SEAT
DANGER
DANGER
RESCUE

SECURITY HEA

PRE STRIKE

QUARTERS, IRAQ
POST STRIKE

AL GHAZAL
100% PURE VEGETABLE GHEE

LIVE
DAVID BLOOM
WITH THE THIRD INFANTRY DIVISION
IN IRAQ
NOW ON MSNBC.COM U.S. STEPS

BA

MISC
FRAGS
AMMO
AMMO

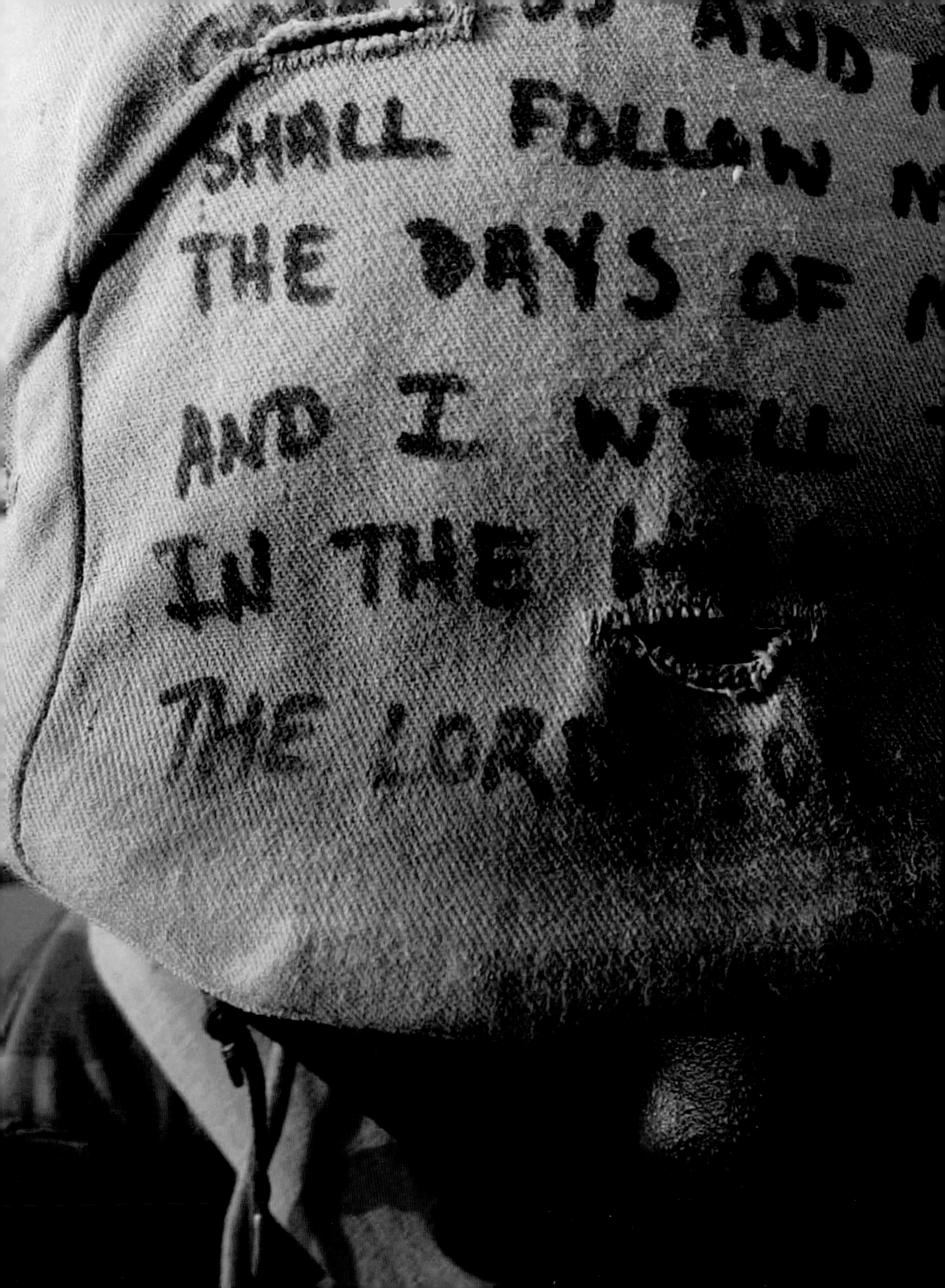
AND
SHALL FOLLOW
THE DAYS OF
AND I
IN THE
THE

OSAUR

الحواسم
السيد الرئيس القائد صدام حسين
يتفقد عددا من مناطق مدينة بغداد

معركة
القائد

Passport control
استلام الامتعة
Baggage claim

TETLEY'S
YOUR KIT

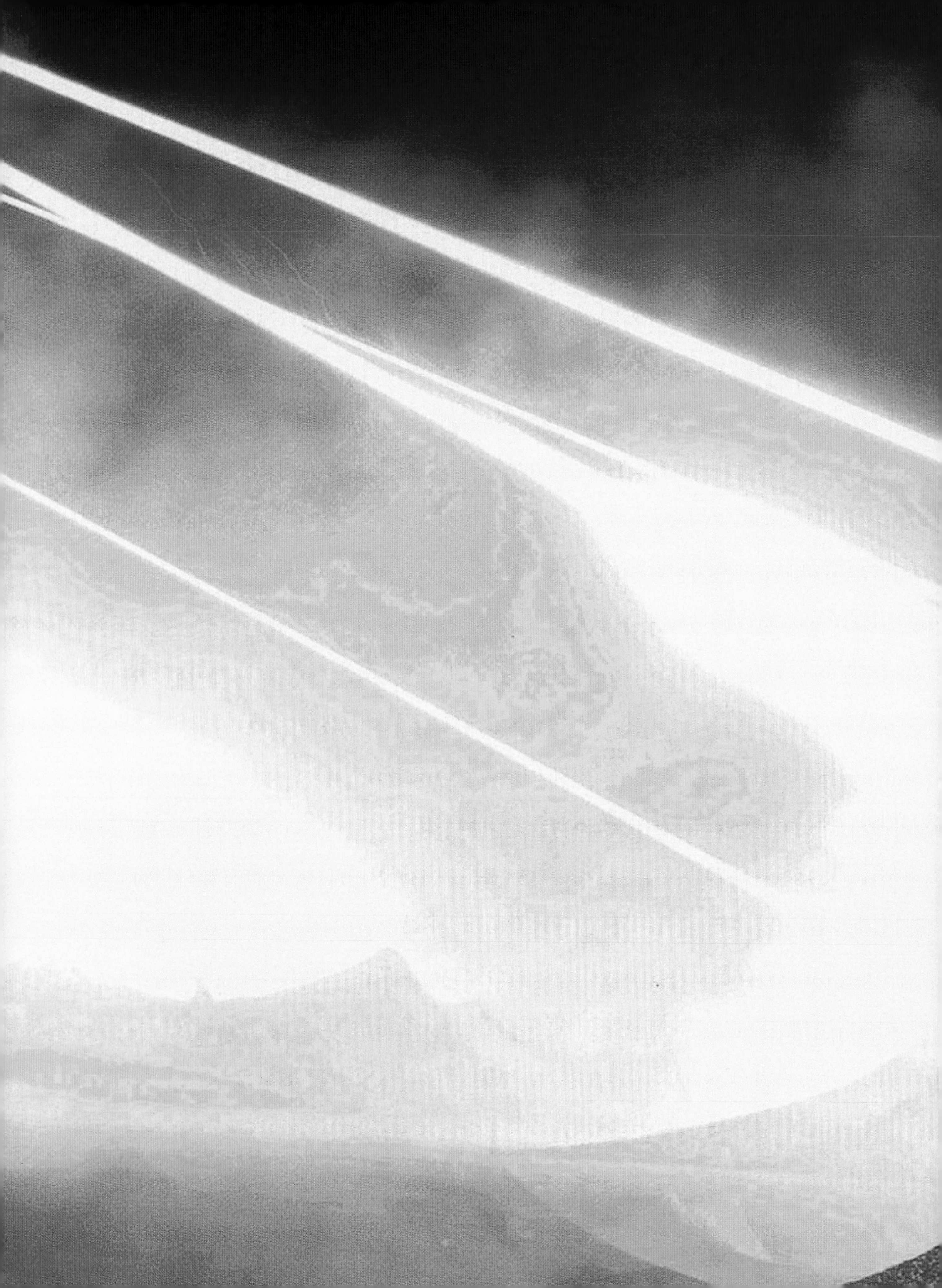

AZTEC
WARRIOR
61

IHA

REPUBLIC OF
MINISTRY
العربية
أبوظبي
LBC
ANN
ARAB NEWS NETWORK
EGYPT
NEWS

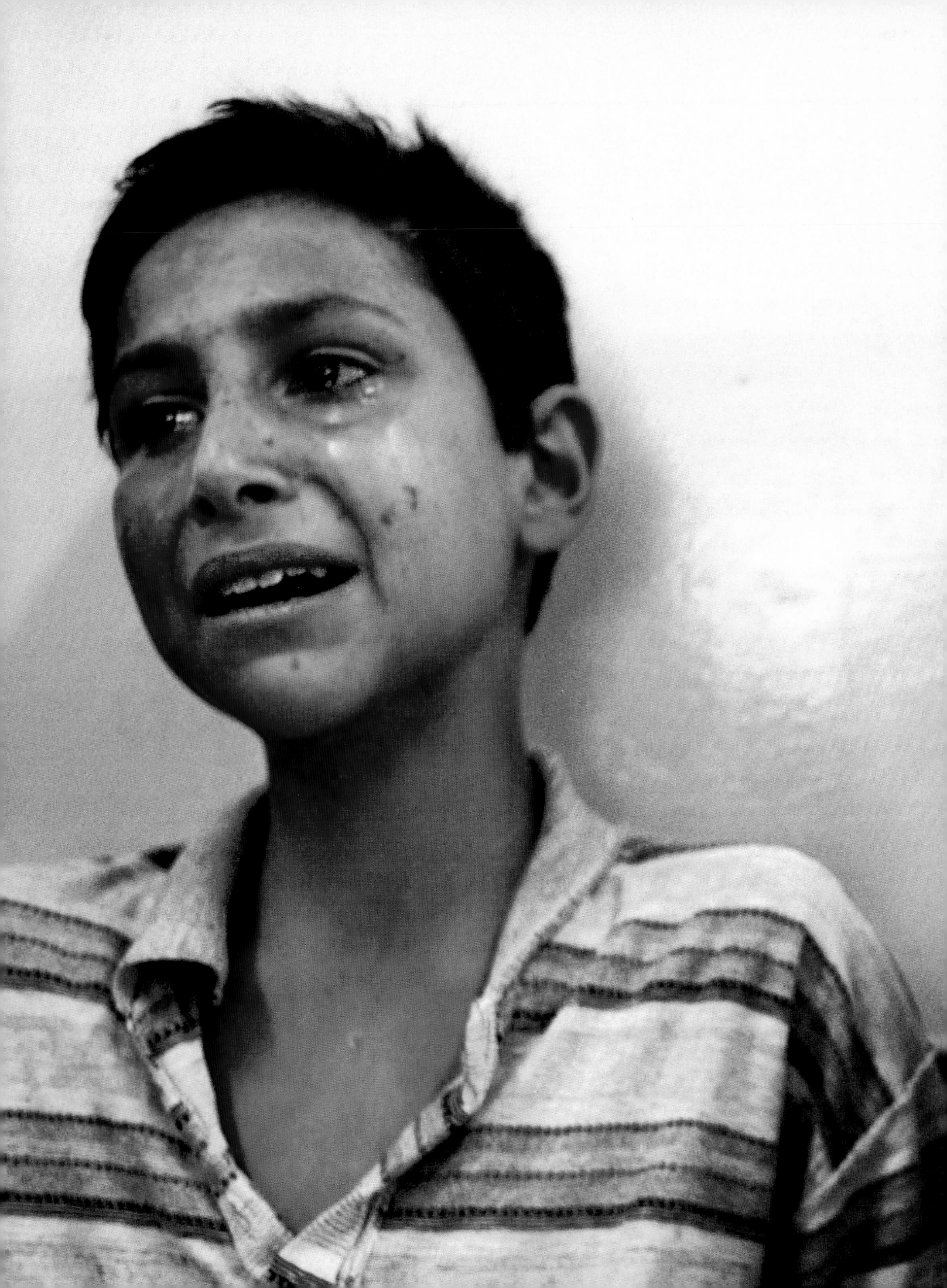

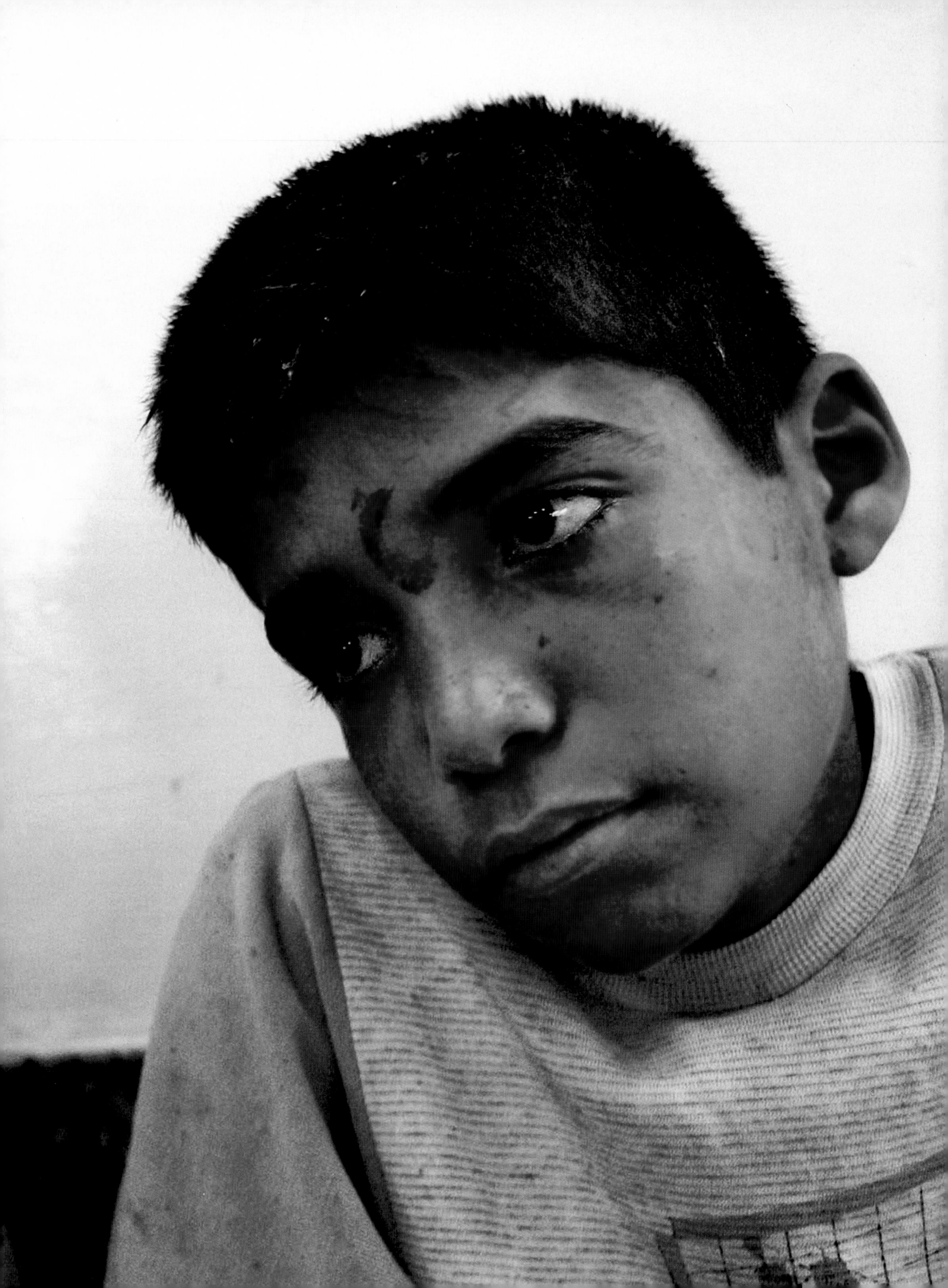

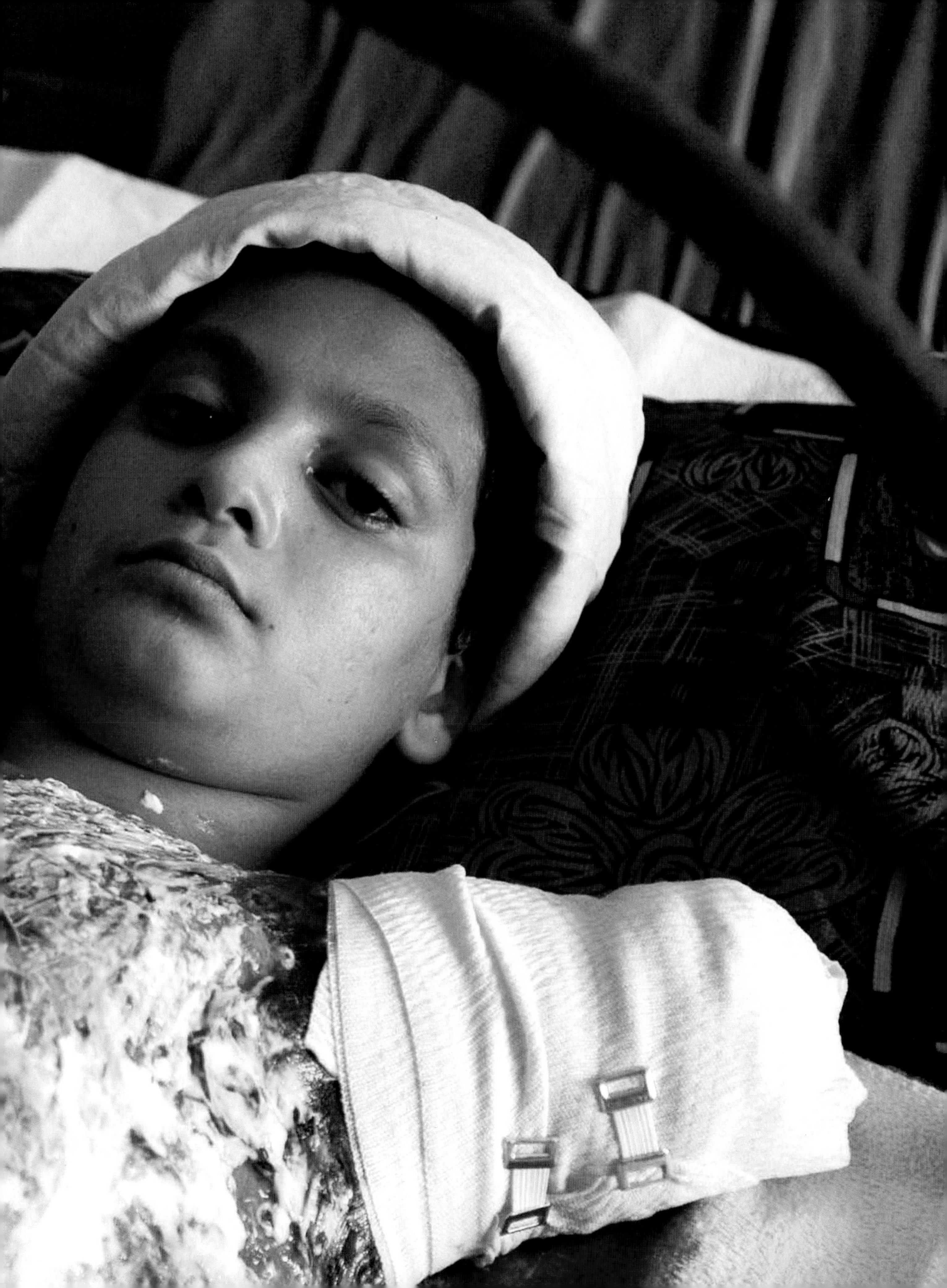

TOP

MACDILL
USAF

محاليل وريدية
شركة

Tahrir Sq.
مستشفى صدام
Saddam Hospi

06

4

SADDAM HUSSEIN'S ART & PHOTO ALBUM

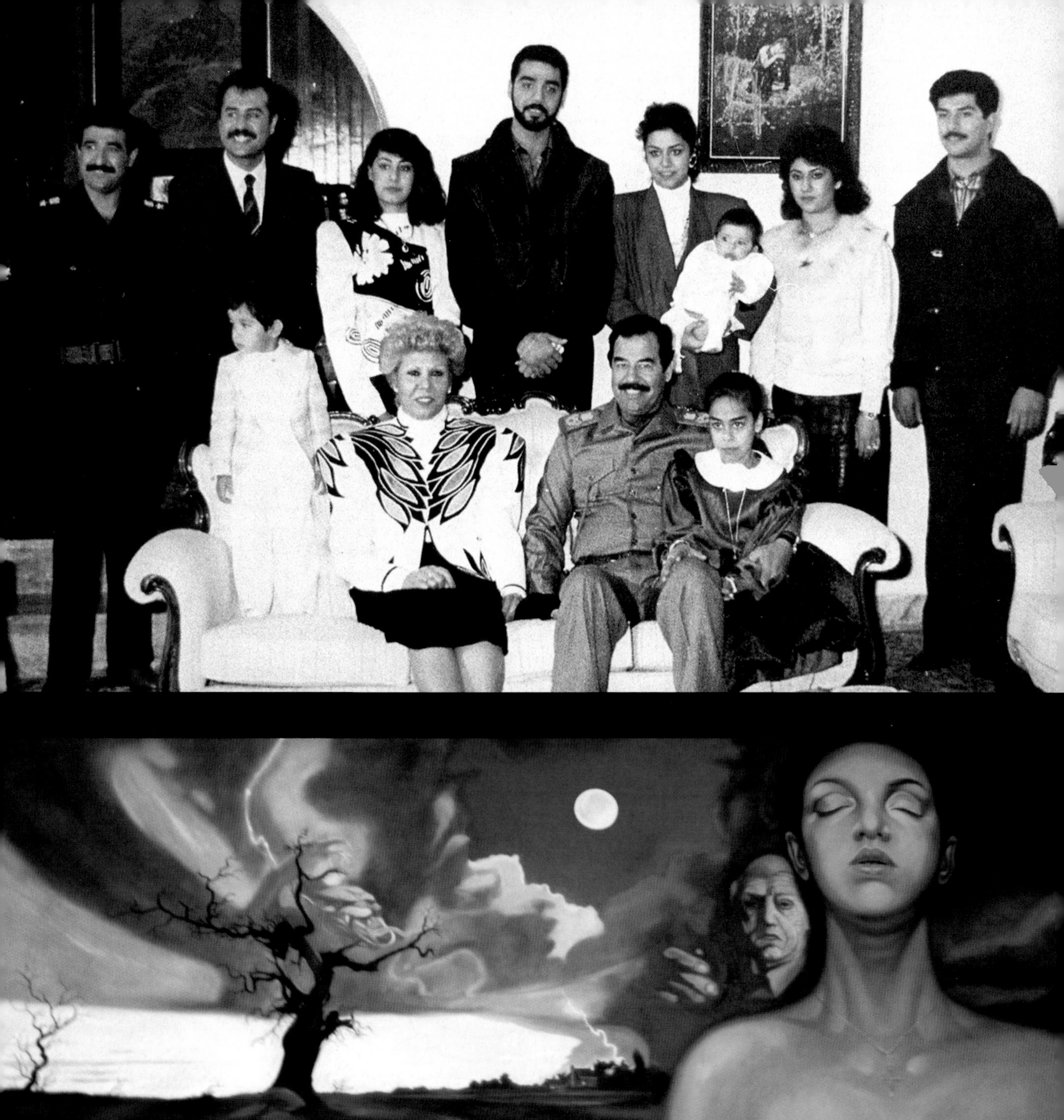

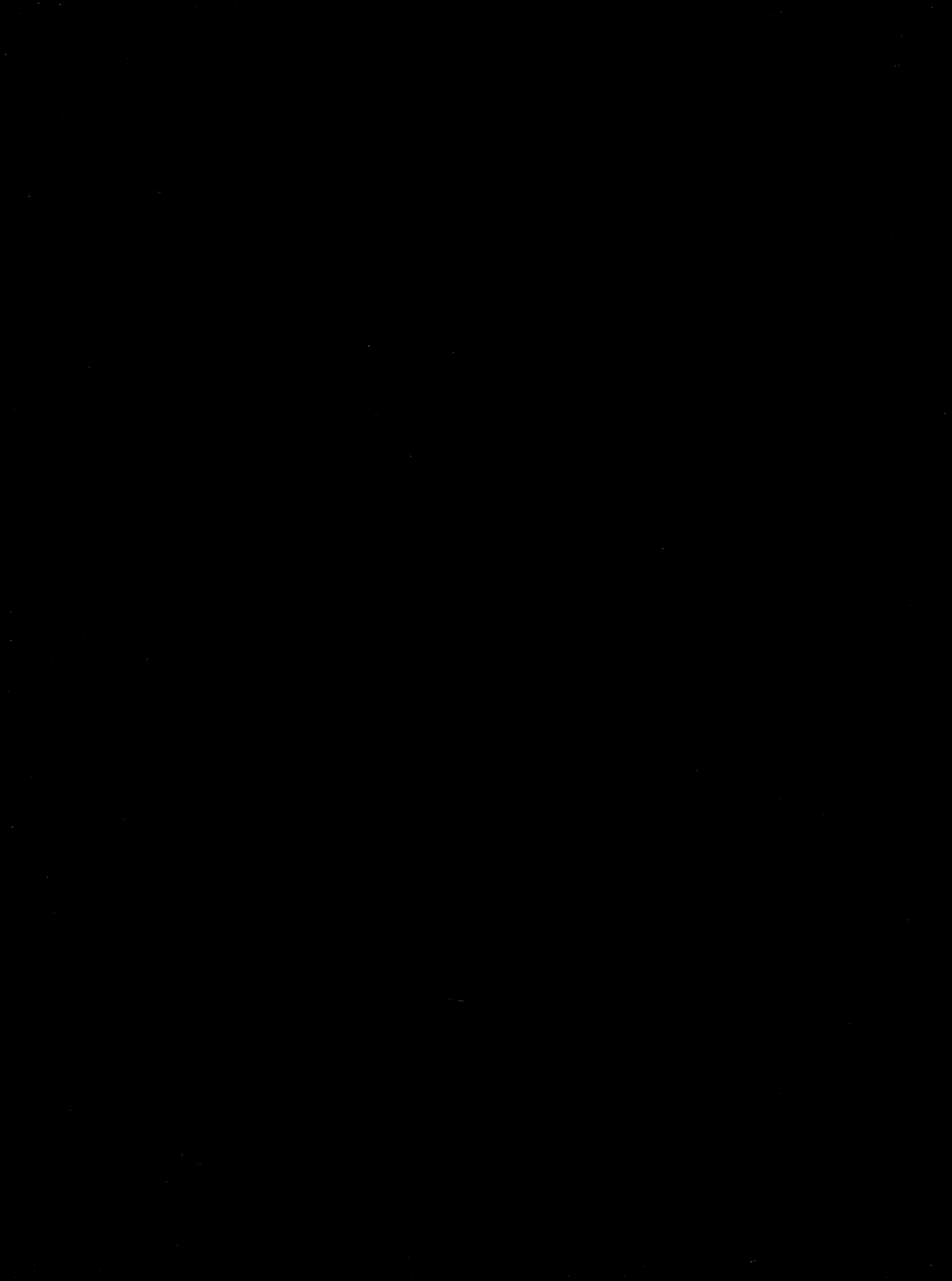

THE FALL OF BAGHDAD

ROMA
mazda

Hi MOM!

الحديثة
TOYOTA
لتجارة زيوت السيارات

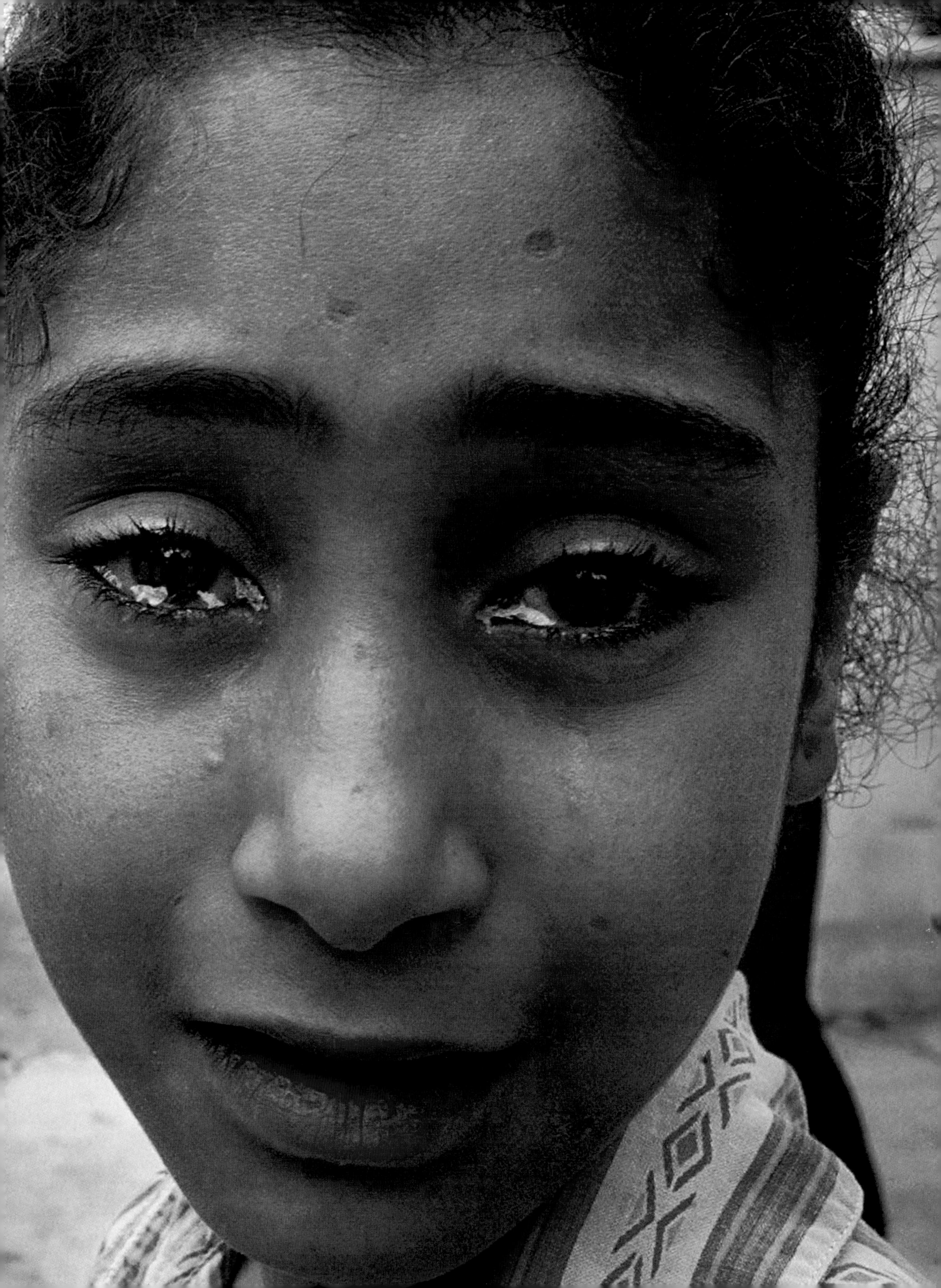

EDUCATION

220V
40W
IRAQ

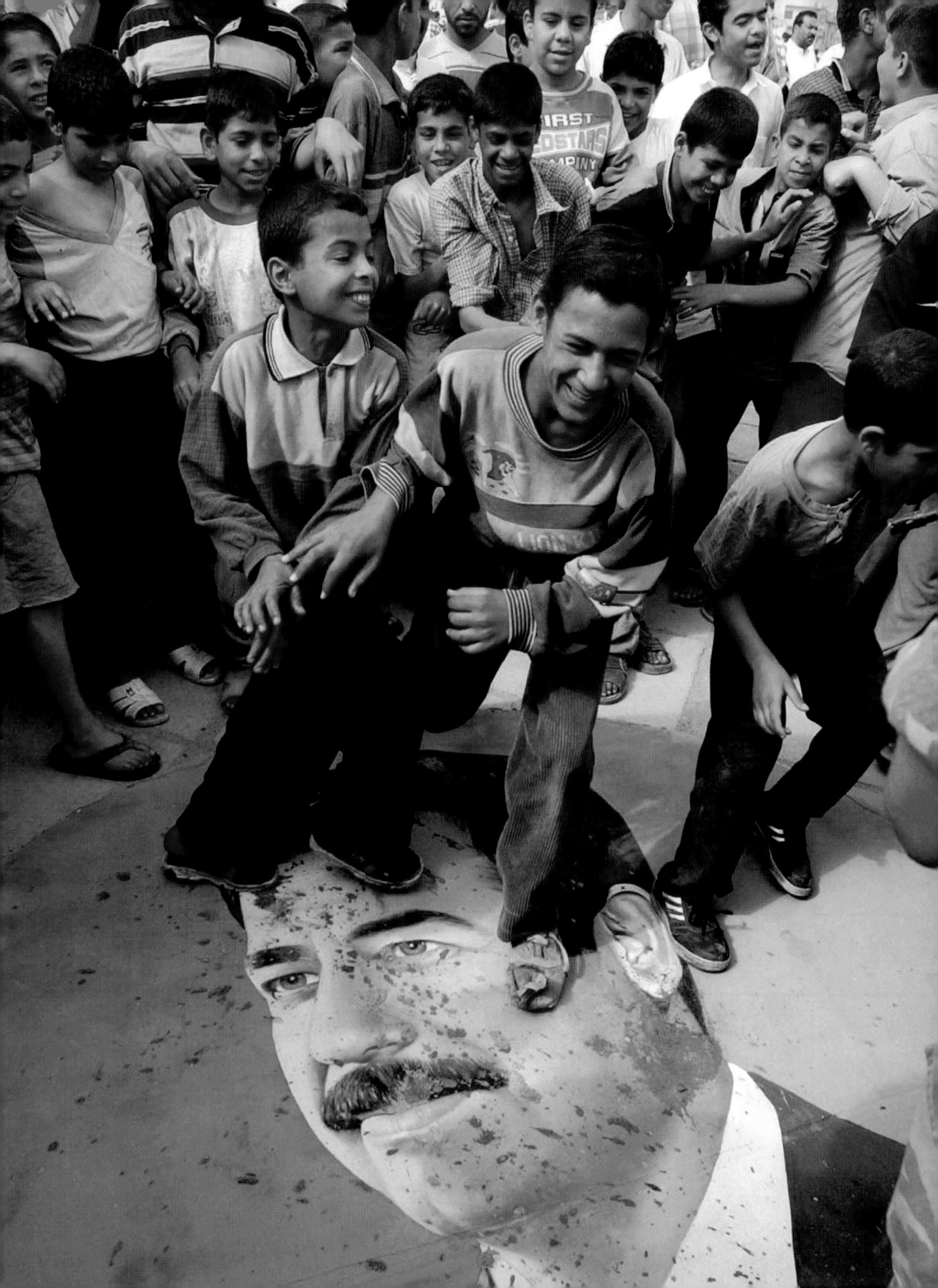
FIRST

KONKA

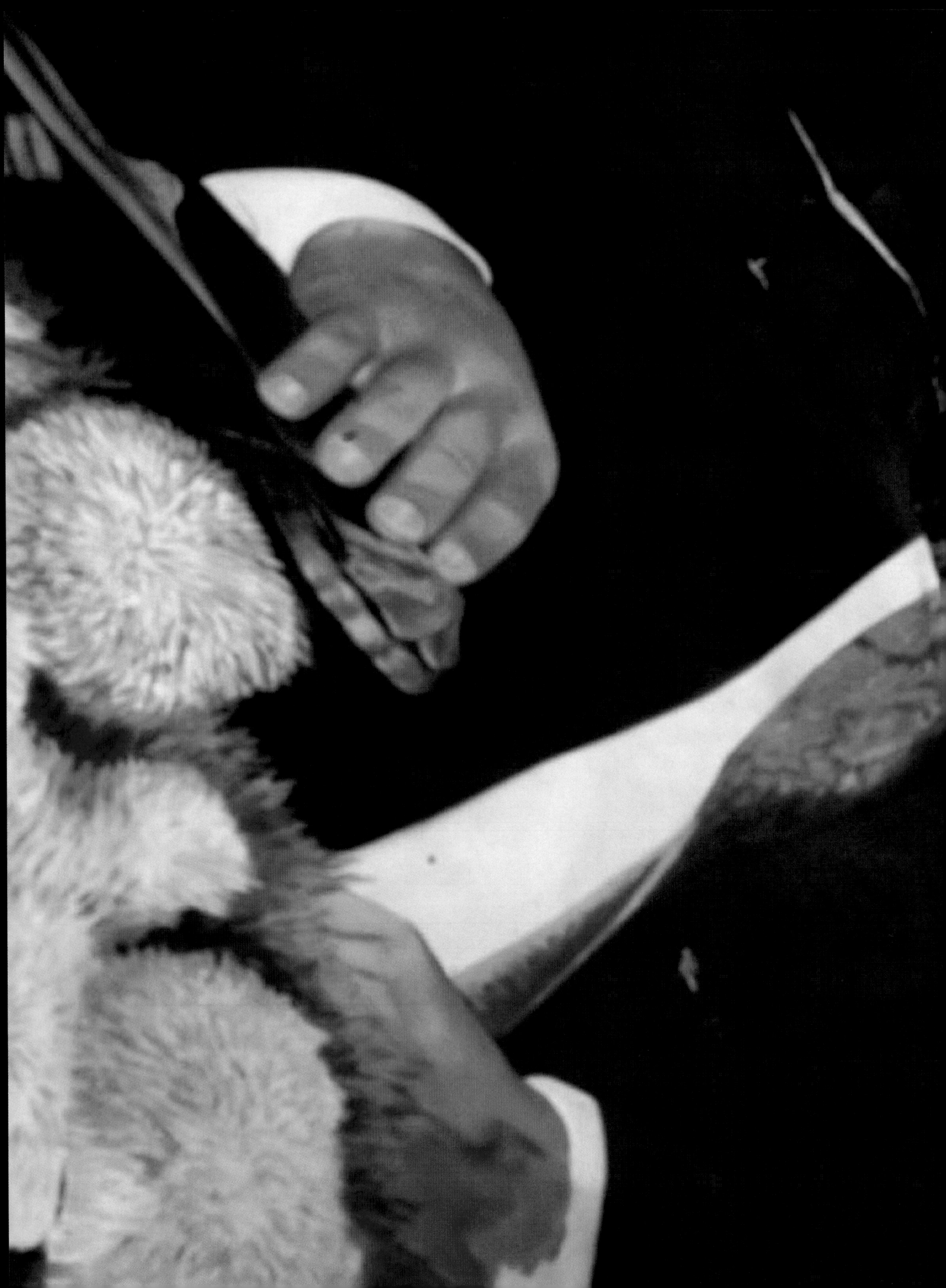

Elite roller

Elite roller

SOUTHERN CEMENT STATE
MADE IN IRAQ

نصر من الله وفتح قريب
القدس لنا
القدس لنا

JOKER
Iraqi Military Ranks
SADDAM HUSAYN AL-TIKRITI
President
ABID HAMID MAHMUD AL-TIKRITI
Presidential Secretary
QUSAY SADDAM HUSAYN AL-TIKRITI
Special Security Organization (SSO) Supervisor/Ba'th Party Military Bureau Deputy Chairman
UDAY SADDAM HUSAYN
National Assembly Member/ Olympic Chairman/ Saddam Fedayeen Chief
ALI HASAN AL-MAJID AL-TIKRITI
Presidential Advisor/ RCC Member
AZIZ SALIH AL-NUMAN
Ba'th Party Regional Command Chairman Responsible for West Baghdad
IZZAT IBRAHIM AL-DURI
RCC Vice Chairman
HANI ABD AL-LATIF TILFAH AL-TIKRITI
SSO Director Special Security Organization
MUHAMMAD HAMZA ZUBAYDI
Retired RCC Member
MUZAHIM SA'B HASAN AL-TIKRITI
Air Defense Forces Commander
KAMAL MUSTAFA ABDALLAH SULTAN AL-TIKRITI
Secretary of the Republican Guard & Special Republican Guard
BARZAN ABD AL-GHAFUR SULAYMAN MAJID AL-TIKRITI
Special Republican Guard Commander
IBRAHIM AHMAD ABD AL-SATTAR MUHAMMAD AL-TIKRITI
Iraqi Armed Forces Chief of Staff
TAHIR JALIL HABBUSH AL-TIKRITI
Iraqi Intelligence Service (IIS)
SAYF AL-DIN FULAYYIH HASAN TAHA AL-RAWI
Republican Guard Chief of Staff
RAFI ABD AL-LATIF TILFAH AL-TIKRITI
Director of General Security (DGS)
HAMID RAJA SHALAH AL-TIKRITI
Air Force Commander
TAHA YASIN RAMADAN AL-JIZRAWI
Vice President/ RCC Member
LATIF NUSAYYIF JASIM AL-DULAYMI
Ba'th Party Military Bureau Deputy Chairman
ABD AL-TAWAB MULLAH HUWAYSH
Deputy Prime Minister
RUKAN RAZUKI ABD AL-GHAFAR SULAYMAN AL-MAJID AL-TIKRITI
Head of Tribal Affairs Office
TAHA MUHYI AL-DIN MARUF
Vice President RCC Member
JAMAL MUSTAFA ABDALLAH SULTAN AL-TIKRITI
Deputy Head of Tribal Affairs Office
MIZBAN KHADR HADI
RCC Member
TARIQ AZIZ
Deputy Prime Minister RCC member
HIKMAT MIZBAN IBRAHIM AL-AZZAWI
Deputy Prime Minister and Finance Minister
WALID HAMID TAWFIQ AL-TIKRITI
Governor of Basrah Governorate
SULTAN HASHIM AHMAD AL-TAI
Minister of Defense
MAHMUD DHIYAB AL-AHMAD
Minister of Interior
AMIR HAMUDI HASAN AL-SADI
Presidential Scientific Advisor
AYAD FUTAYYIH KHALIFA AL-RAWI
Quds Forces Chief of Staff
ZUHAYR TALIB ABD AL-SATTAR AL-NAQIB
Director of Military Intelligence (DMI)
AMIR RASHID MUHAMMAD AL-UBAYDI
Presidential Advisor/ Oil Minister
SABAWI IBRAHIM HASAN AL-TIKRITI
Presidential Advisor
HUSAM MUHAMMAD AMIN AL-YASIN
Head of National Monitoring Directorate (NMD)
MUHAMMAD MAHDI AL-SALIH
Minister of Trade
WATBAN IBRAHIM HASAN AL-TIKRITI
Presidential Advisor
ABD AL-BAQI ABD AL-KARIM ABDALLAH AL-SADUN
Ba'th Party Regional Command Chairman for Diyala District
BARZAN IBRAHIM HASAN AL-TIKRITI
Presidential Advisor
HUDA SALIH MAHDI AMMASH
WMD Scientist/ Ba'th Party Regional Command Member
MUHAMMAD ZIMAM ABD AL-RAZZAQ AL-SADUN
Ba'th Party Regional Command Chairman for Ta'mim Governorate
YAHYA ABDALLAH AL-UBAYDI
Ba'th Party Regional Command Chairman for Basrah Governorate
SAMIR ABD AL-AZIZ AL-NAJIM
Ba'th Party Regional Command Chairman for East Baghdad Governorate
HUMAM ABD AL-KHALIQ ABD AL-GHAFUR
Minister of Higher Education and Scientific Research
SA'D ABDUL-MAJID AL-FAISAL AL-TIKRITI
Ba'th Party Regional Command Chairman for Salah Al-Din Governorate
MUHSIN KHADR AL-KHAFAJI
Ba'th Party Regional Command Chairman for Qadisiyah District
SAYF AL-DIN AL-MASHHADANI
Ba'th Party Regional Command Chairman for Muthanna District
FADIL MAHMUD GHARIB (AKA: GHARIB MUHAMMAD FAZEL AL-MASHAIKH)
Ba'th Party Regional Command Chairman for Babil District
RASHID TAAN KAZIM
Ba'th Party Regional Chairman for Al-Anbar Governorate
ADIL ABDALLAH MAHDI
Ba'th Party Regional Command Chairman for Dhi Qar District
UGLA ABID SAQR AL-KUBAYSI
Ba'th Party Regional Chairman for Maysan Governorate
GHAZI HAMMUD AL-UBAYDI
Ba'th Party Regional Command Chairman for Al-Kut District
JOKER
Arab Titles

US MILITARY

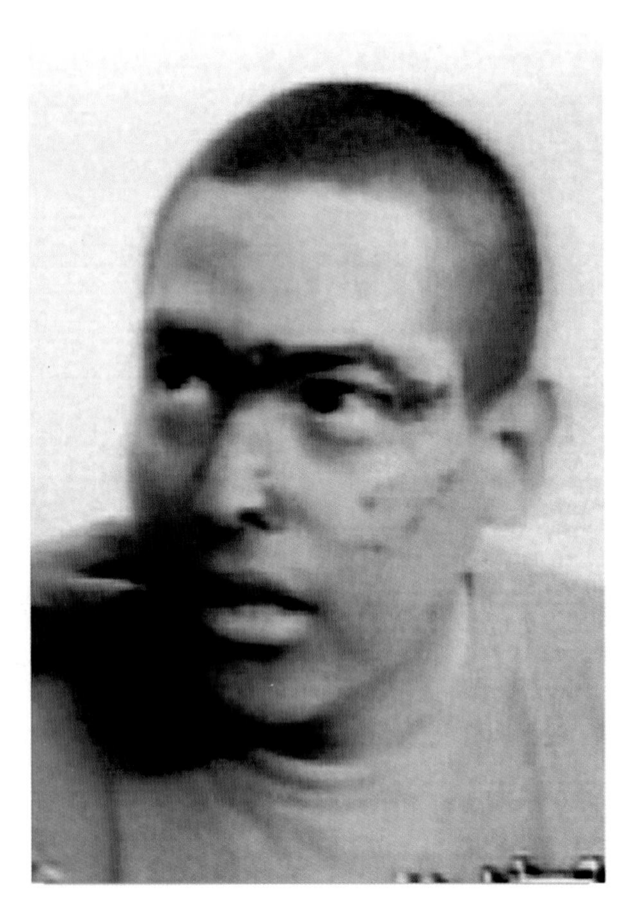

Edgar Hernandez

Army Spc.
Joseph Hudson

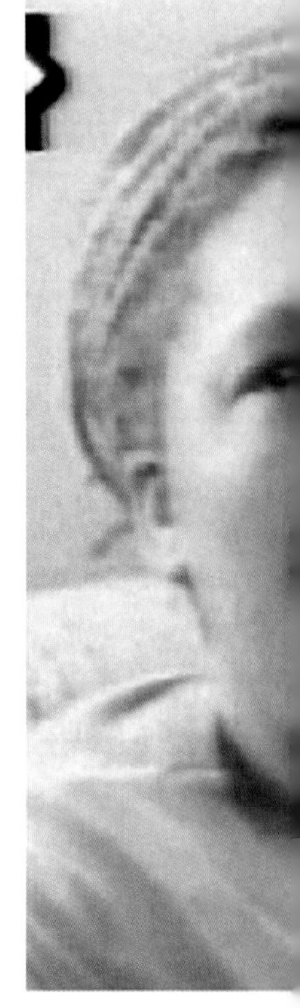

Arm
Shoshaw

RAQ WAR POW

Spc.
Johnson

Pfc. Patrick Miller

Sgt. James Riley

US

AIR FORCE

Thank You

Bosh and Blear

ii–iii. *Camp Pennsylvania, Kuwait, March 15, 2003:* A soldier with the 101st Airborne Division during a training operation.
Benjamin Lowy/Corbis

iv–v. *USS* Theodore Roosevelt, *Eastern Mediterranean Sea, March 10, 2003:* A fighter leaves the deck at sunset.
Findlay Kember/Polaris

vi–vii. *Southern Iraqi desert, March 28, 2003:* Marines from the 3rd Battalion hold their positions on the road to Baghdad.
Kuni Takahashi/*Boston Herald*/ReflexNews

viii–ix. *Southern Iraqi desert, March 22, 2003:* A Marine takes cover as Iraqis fire flares into the early morning sky.
Kuni Takahashi/*Boston Herald*/ReflexNews

THE ROAD TO WAR

2–3. *Najaf, Iraq, April 7, 2003:* General Tommy Franks (left), commander of the U.S. forces in Iraq, addresses soldiers from the 101st Airborne Division.
Benjamin Lowy/Corbis

4–5. *Washington, D.C., October 16, 2002:* President George W. Bush signs the congressional resolution authorizing use of force against Iraq.
Chris Corder/UPI Photo Service

6–7. *Baghdad, September 16, 2002:* Civilians walk past a mural of Iraqi president Saddam Hussein on horseback.
Amr Nabil/AP Wide World Photos

8. *New York, March 17, 2003:* United Nations Secretary General Kofi Annan answers questions from the media.
Jason Szenes/Corbis Sygma

9. *Baghdad, January 18, 2003:* Iraqis protest the presence of United Nations weapons inspectors outside United Nations headquarters in the capital.
Hussein Malla/AP Wide World Photos

10. (top) *Baghdad, October 2002:* Saddam Hussein praised on Iraqi television.
Thomas Dworzak/Magnum Photos

10. (bottom) *Baghdad, October 2002:* Memorial commemorating the anniversary of a school being bombed on the outskirts of Baghdad during the Iran-Iraq war.
Thomas Dworzak/Magnum Photos

11. *Baghdad, March 26, 2003:* Outdoor market during a sandstorm.
Jerome Delay/AP Wide World Photos

12–13. *Baghdad, February 20, 2003:* Iraqi men in a café.
Bruno Stevens/Cosmos/Aurorqa

14–15. *Lajes, Portugal, March 16, 2003:* President George W. Bush at a press conference.
Lionel Cironneau/AP Wide World Photos

16–17. *As Sayliyah Base, Qatar, December 12, 2002:* Secretary of Defense Donald Rumsfeld (right) and General Tommy Franks address members of the U.S. military.
Karen Ballard/Corbis Sygma

18. *New York, February 5, 2003:* U.S. Secretary of State Colin Powell holds a vial that he described as one that could contain anthrax during his presentation on Iraq to the United Nations Security Council.
UN/DPI photo by Mark Garten/Corbis Sygma

19. *New York, February 5, 2003:* Iraqi Ambassador to the United Nations Mohammed Al-Duri reacts to Secretary Powell's presentation to the United Nations.
Lorenzo Ciniglio/Corbis Sygma

20–21. *Washington, D.C., February 5, 2003:* President George W. Bush watches the broadcast of Secretary Powell's presentation to the United Nations.
Eric Draper/White House Photo/Corbis Sygma

22–23. *USS* Abraham Lincoln, *Arabian Sea, February 12, 2003:* A Navy flight deck attendant preps an E-2C Hawkeye for takeoff.
Adam Butler/AP Wide World Photos

24–25. *Fayetteville, North Carolina, February 13, 2003:* Brigade General Abe Turner addresses members of the 82nd Airborne, 2nd Brigade Combat Team, before deployment to Kuwait.
Ellen Ozier/Reuters

26. *Baghdad, date unknown:* Muslim volunteer prepares to fight.
Markus Matzel/Black Star

27. *Baghdad, date unknown:* Iraqi forces prepare to fight.
Markus Matzel/Black Star

28. *USS* Abraham Lincoln, *February 21, 2003:* Flight crew members talk as an F-14 Tomcat takes off.
Leila Gorchev/AP Wide World Photos

29. *Baghdad, July 12, 2001:* Young militia volunteers.
Karim Mohsen/Getty Images

30–31. *Kuwaiti desert, December 20, 2002:* U.S. military vehicles take part in exercises.
Anja Niedringhaus/AP Wide World Photos

32. *Baghdad, March 5, 2003:* Iraqi soldiers brandish weapons at a military parade.
Naim Assaker/Corbis

33. *Killeen, Texas, March 27, 2003:* Soldiers with the U.S. Army 4th Infantry Division say good-bye to friends and family before deployment to Iraq.
Brad Loper/*Dallas Morning News*/Corbis

34. *USS* Abraham Lincoln, *Arabian Sea, March 21, 2003:* Skids of ordnance wait in the hangar bay to be transferred to the flight deck.
Jason Frost/AP Wide World Photos

35. *Fort Campbell, Kentucky, December 2, 2003:* Staff Sergeant Calvin Coats demonstrates camouflage paint application to members of the 101st Airborne Division.
John Sommers/Reuters

36. *USS* Abraham Lincoln, *Arabian Sea, February 11, 2003:* A weapons and ammunition handler sits on the wing of a fighter.
Adam Butler/AP Wide World Photos

37. *Fort Benning, Georgia, January 6, 2003:* Cody Sochocki, six, says good-bye to his father, Staff Sergeant John Sochocki, as the 3rd Brigade, 3rd Infantry, prepares to deploy to the Middle East.
Robin Trimarchi/*Columbus Ledger-Enquirer*/Corbis Sygma

38–39. *Baghdad, February 6, 2003:* An Iraqi fisherman prays by the Tigris River.
Bruno Stevens/Cosmos/Aurora

40–41. *St. Petersburg, Russia, February 14, 2003:* Russian matryoshka dolls on display.
Alexander Demianchuk/Reuters

42–43. *New York, March 23, 2003:* Demonstrators rally in Times Square in support of President Bush and American troops.
Susan Meiselas/Magnum Photos

44. (far left) *New York, February 15, 2003:* Antiwar march.
Diane Bondareff/AP Wide World Photos

44. (middle) *Madrid, February 15, 2003:* Antiwar march.
Paul White/AP Wide World Photos

44–45. *Paris, February 15, 2003:* Antiwar march.
Laurent Rebours/AP Wide World Photos

45. (middle) *London, February 15, 2003:* Antiwar march.
Andrew Parsons/AP Wide World Photos

45. (far right) *Berlin, February 15, 2003:* Antiwar march.
Jan Bauer/AP Wide World Photos

46. (top) *Washington, D.C., January 18, 2003:* Antiwar rally.
Jason Turner/AP Wide World Photos

46. (middle) *Edinburgh, Scotland, February 14, 2003:* An antiwar protester holds up a banner as British prime minister Tony Blair arrives at the Royal Infirmary.
Jeff J. Mitchell/Reuters

46. (bottom) *Madras, India, February 14, 2003:* Students at a rally aimed at promoting peace.
L. Anatha Krishan/Reuters

7. *Baghdad, October 17, 2002:* Saddam Hussein wields a sword given to him before being sworn in as president for the next seven years.
AP Wide World Photos/INA

8. (top) *Berlin, February 24, 2003:* French president Jacques Chirac (left) and German chancellor Gerhard Schroeder address the media.
Markus Schreiber/AP Wide World Photos

8. (bottom) *St. Petersburg, April 11, 2003:* (from left) French president Jacques Chirac, Russian president Vladimir Putin, and German chancellor Gerhard Schroeder at a joint press conference.
Remy de la Mauviniere/AP Wide World Photos

9. *Terceira, Portugal, March 16, 2003:* (from left) British prime minister Tony Blair, President George Bush, and Spanish prime minister José Maria Anzar after their arrival at the Azores islands.
Kieran Doherty/Reuters

0. *Baghdad, 1985:* Saddam Hussein wristwatch.
Steve McCurry/Magnum Photos

1. (top) *Washington, D.C., March 19, 2003:* President George W. Bush announces that he has ordered war to disarm Iraq.
CNN/Agence France Presse

1. (bottom) *Hebron, Israel, March 20, 2003:* A Palestinian family watches a speech by Saddam Hussein that aired three hours after the beginning of Operation Iraqi Freedom.
Nayef Hashlamoun/Reuters

2–53. *Camp New York, Kuwait, no date available:* Chow line in the desert.
Bassignac Gilles/Gamma

4. *USS* Carl Vinson, *February 9, 2003:* An aircraft director guides an F/A-18C Hornet onto a steam-powered catapult during training exercises in the Pacific.
U.S. Navy

5. *USS* Bunker Hill, *Persian Gulf, March 19, 2003:* Tomahawk Land Attack Missile launch.
U.S. Navy

6–57. *Baghdad, March 20, 2003:* Baghdad explosion at dawn.
ITAR-TASS Photos

8–59. *Meridian, Texas, March 20, 2003:* The Winkler family watches coverage of the bombing of Baghdad from their family farm.
Steve Liss/Corbis

0–61. *Baghdad, March 21, 2003:* The city burns after an allied bombing strike.
Mike Moore/*Daily Mirror*/Corbis

2–63. *Baghdad, March 21, 2003:* Explosion at one of Saddam Hussein's palaces.
Ilkka Uimonen/Magnum Photos

VAR

6–67. *North Rameila, Iraq, March 25, 2003:* Members of British 16 Air Assault Brigade, 3 Army Air Corps, on patrol during a sandstorm.
Ian Jones/AP Wide World Photos

8–69. *Outskirts of Nasiriyah, Iraq, March 25, 2003:* Marines from the 2nd Battalion, 8th Regiment, prepare to enter an Iraqi military hospital.
Chris Bouroncle/Agence France Presse

0–71. *Outside Karbala, Iraq, March 26, 2003:* An Iraqi boy and small child ride a camel through a sandstorm.
Peter Andrews/Reuters

2–73. *Location unknown, March 26, 2003:* Airmen walk through a sandstorm at a forward-deployed air base.
Corbis

4–75. *Najaf, Kuwait, March 26, 2003:* A U.S. Army combat engineer smokes during a sandstorm.
Kai Pfaffenbach/Reuters

6–77. *Iraqi desert, March 25, 2003:* An Iraqi man surrenders to U.S. troops during a sandstorm.
David Leeson/*Dallas Morning News*/Corbis

78–79. *Iraqi desert, March 25, 2003:* Sgt. Aaron Knizley on guard duty in a sandstorm.
Kuni Takahashi/*Boston Herald*/ReflexNews

80–81. *Baghdad, March 27, 2003:* Iraqi civilians in a sandstorm.
Jobard/Sipa

82–83. *Nasiriyah, Iraq, March 26, 2003:* U.S. Marines wake up in a mudfield after heavy sandstorms.
Eric Feferberg/Agence France Presse

84–85. *Al Sayliyah Base, Qatar, March 22, 2003:* General Tommy Franks addresses members of the press at the U.S. CENTCOM headquarters.
Karen Ballard/Corbis Sygma

86–87. *Basra, Iraq, April 14, 2003:* A British Army Challenger II tank and two of its crew at an operational post.
Daily Mirror/Getty Images

88. *Baghdad, March 26, 2003:* An Iraqi woman cries after learning of the death of her brother, a civilian.
Bruno Stevens/Cosmos/Aurora

89. *Al Hilla, Iraq, April 1, 2003:* Slain Iraqi woman and child in a coffin.
Akram Saleh/Reuters

90–91. *Outside Basra, Iraq, March 28, 2003:* British soldiers man the last checkpoint before Basra.
Paolo Pellegrin/Magnum Photos

92–93. *Al Faysaliyah, Iraq, March 25, 2003:* PFC Joseph Dwyer carries an Iraqi boy who was injured during a heavy battle outside the village while the boy's father (far left) goes back to help others.
Warren Zinn/*Army Times*/Corbis

94–95. *McAlester, Oklahoma, March 28, 2003:* Carla Bagley looks at a picture of her son, Clint, a sailor aboard the aircraft carrier USS *Abraham Lincoln*, stationed in the Arabian Sea.
Kevin Harvison/ AP Wide World Photos

96–97. *Rumaila oil fields, Southern Iraq, March 17, 2003:* Crews battle oil well blazes set by Iraqi forces.
Hyoung Chang/*Denver Post*/Polaris

98–99. *Baghdad, March 27, 2003:* Iraqi child's artwork in an orphanage.
Jobard/Sipa

100. *Baghdad, March 28, 2003:* Iraqis pray over the remains of family members.
Jerome Delay/AP Wide World Photos

101. *Basra, Iraq, March 28, 2003:* An Iraqi family flees past a destroyed T-55 tank after a mortar attack on British army positions.
Chris Helgren/Reuters

102–103. *Baghdad, March 28, 2003:* Smoke from burning oil billows around a statue of Saddam Hussein.
Ilkka Uimonen/Magnum Photos

104–105. *Qhura Anjir, Iraq, March 28, 2003:* A tractor destroys a mural of Saddam Hussein after Kurdish soldiers secured the area.
Newsha Tavakolian/AP Wide World Photos

106. *Qhura Anjir, Iraq, March 28, 2003:* Destroyed mural of Saddam Hussein.
Newsha Tavakolian/Polaris

107. *Southern outskirts of Baghdad, April 10, 2003:* A local office of the Baath Party burns.
Goran Tomasevic/Reuters

108–111. *Az Zubaya, Iraq, March 29, 2003:* Members of the British Light Infantry distribute aid packages to Iraqi civilians outside of Basra.
Brian Roberts/AP Wide World Photos

112–113. *Basra, Iraq, March 30, 2003:* Iraqi civilians flee the besieged city.
Paolo Pellegrin/Magnum Photos

114. *Az Zubair, Iraq, March 27, 2003:* A British army Challenger II tank crushes a portrait of Saddam Hussein outside of Basra.
Chris Helgren/Reuters

115. *Basra, Iraq, March 30, 2003:* An Iraqi girl looks up at a British soldier who is reviewing her father's papers at a checkpoint.
Odd Andersen/Agence France Presse

116–117. *Baghdad, March 30, 2003:* An Iraqi man sits amid the remains of his house, destroyed during an allied airstrike on a nearby target.
Jobard/Sipa

118–119. *Basra, Iraq, March 29, 2003:* An Iraqi girl holds her sister as their mother buys food.
Jerry Lampen/Reuters

120–121. *Basra, Iraq, March 29, 2003:* An Iraqi man and child walk on a road outside of the besieged city as oil fires burn in the distance.
Spencer Platt/Getty Images

122. *Southern Iraqi desert, April 4, 2003:* An A-10 Warthog lands at a forward coalition air base.
Tim Sloan/Agence France Presse

123. *Southern Iraqi desert, March 31, 2003:* A U.S. Army guard at an allied air base.
Tim Sloan/Agence France Presse

124–125. *Iraqi desert, March 30, 2003:* Inside the Command Post at the Headquarters of 3 Commando Brigade.
Simon Walker/*The Times*/News International Photos

126. *Umm Qasr, Iraq, March 31, 2003:* An Iraqi boy rides his bike past a defaced mural of Saddam Hussein.
Jon Mills/AP Wide World Photos

127. *Sulaymaniyah City, Iraq, March 31, 2003:* A Peshmerga fighter in an Iraqi hospital with a depiction of the New York skyline hanging above his bed.
Francesco Zizola/Magnum Photos

128. *Kuwait, March 20, 2003:* A U.S. Marine in a gas mask and chemical protection suit after a warning of Iraqi missile launches was sent out.
Russell Boyce/AP Wide World Photos

129. *Az Zubaya, Iraq, April 1, 2003:* British Lance Corporal Graeme Church sits on the head of a destroyed statue of Saddam Hussein.
Brian Roberts/AP Wide World Photos

130–131. *Basra, Iraq, April 1, 2003:* A soldier from the Irish Guard 7th Armour Brigade searches a man on the way out of Basra.
Mario Tama/Getty Images

132–133. *Iraq, April 1, 2003:* Night-vision footage of U.S. Army Private First Class Jessica Lynch's rescue.
Corbis

134–135. An undated photo of U.S. Army PFC Jessica Lynch, who was captured by Iraqi forces but later rescued.
Handout/Agence France Presse

136–137. *Baghdad, March 22, 2003:* Smoke billows from oil fires in and around the Iraqi capital.
Mike Moore/*Daily Mirror*/Mirrorpix

138–139. *Baghdad, March 23, 2003:* Smoke billows from oil fires in and around the Iraqi capital.
Jobard/Sipa

140–141. *Unknown location and date:* U.S. Air Force pararescue men perform a 12,999 foot High Altitude Low Opening free-fall drop from a C-130 Hercules during a training mission above a forward deployed Operation Enduring Freedom base.
Jeremy T. Lock/U.S. Air Force

142–143. *Outside Kuwait City, Kuwait, March 11, 2003:* British soldiers of the 51st Squadron Royal Air Force regiment watch the airlifting of a military vehicle during a training operation.
Laurent Rebours/ AP Wide World Photos

144–145. *Diala, Iraq, January 7, 2003:* Iraqi women army volunteers march during a military parade.
Corbis Sygma

146–147. *Camp Shoup, Kuwait, March 20, 2003:* U.S. Marines prepare before crossing the border into Iraq.
Eric Feferberg/Agence France Presse

148. *Outskirts of Baghdad, April 6, 2003:* A U.S. Marine surveys the streets during a fight for the Baghdad Highway Bridge.
Kuni Takahashi/*Boston Herald*/ReflexNews

149. *USS* Gunston Hall, *Arabian Gulf, March 18, 2003:* A bottlenose dolphin belonging to the Commander Task Unit leaps out of the water during a training exercise.
Brien Aho/Reuters

150–151. *U.S. air base, Arabian Gulf, March 14, 2003:* A-10 Warthogs prepare for takeoff.
Paula Bronstein/Getty Images

152–153. *Released April 1, 2003 by U.S. Central Command:* Pre- and post-strike views of a security headquarters compound in Iraq.
AP Wide World Photos/Department of Defense/Handout

154–155. *Southern Iraqi desert, March 27, 2003:* Allied convoy on the road to Baghdad.
Romeo Gacad/Corbis/Agence France Presse

156–157. *Safwan, Iraq, March 21, 2003:* Iraqi soldiers surrender to U.S. Marines.
Christophe Calais/*In Visu*/Corbis

158–159. *Karbala, Iraq, March 27, 2003:* U.S. Army M1-A1 tanks move into position in central Iraq.
John Moore/AP Wide World Photos

160–161. *Washington, D.C., January 6, 2003:* (from left) Secretary of the Interior Gail Norton, Secretary of State Colin Powell, President George W. Bush, and Secretary of Defense Donald Rumsfeld hold a cabinet meeting at the White House.
Brooks Kraft/Corbis

162–163. *Baghdad, April 10, 2003:* Relatives grieve after a father, son, and another man were shot after failing to stop when ordered to by Marines.
Ilkka Uimonen/Magnum Photos

164–166. *Nasiriyah, Iraq, March 26, 2003:* A U.S. Marine from Task Force Tarawa takes in surrendering Iraqis.
Joe Raedle/Getty Images

167. *Basra, Iraq, April 4, 2003:* Iraqi women reach out with empty containers as British soldiers arrive with a supply of fresh water.
Anja Niedringhaus/AP Wide World Photos

168. *Karbala, Iraq, April 3, 2003:* A U.S. Army combat crew member mans the gun of a Bradley armored vehicle.
Romeo Gacad/Agence France Presse

169. *Baghdad, April 3, 2003:* A member of the Iraqi Republican Guard on the outskirts of the capital.
Patrick Baz/Agence France Presse/Corbis

170–171. *Karbala, Iraq, April 2, 2003:* An Iraqi man reaches out and kisses Private Garrett Alvin at a U.S. Army checkpoint.
David Leeson/*Dallas Morning News*/Corbis

172–173. *Karbala, Iraq, April 3, 2003:* Secondary explosions from a destroyed Iraqi SA-6 surface-to-air missile shoot across the sky.
David Leeson/*Dallas Morning News*/Corbis

174. *Iraqi desert, 2003:* NBC News correspondent David Bloom reports from Iraq.
NBC News/AP Wide World Photos

175. *St. Anne, Illinois, April 3, 2003:* A mourner stands in front of the casket of U.S. Marine Captain Ryan Beaupre, who was killed on March 21 when his helicopter crashed in Kuwait.
Scott Olson/Getty Images

176–177. *Basra, Iraq, April 3, 2003:* Iraqi men remove the body of an Iraqi soldier who was killed during the battle for Basra.
(176 top) Paolo Pellegrin/Magnum Photos
(176 bottom) Mario Tama/Getty Images
(177) Peter Turnley/*Denver Post*/Corbis

178–179. *Kuwaiti desert, March 18, 2003:* A U.S. Army soldier walks past a line of Bradley armored vehicles.
John Moore/AP

180–181. *Basra, Iraq, April 4, 2003:* Allied soldiers hold their positions during the push through Basra.
Paolo Pellegrin/Magnum Photos

82–183. *Najaf, Kuwait, April 4, 2003:* A U.S. Army combat engineer, with Psalm 21 written on his helmet, takes a short rest.
Kai Pfaffenbach/Reuters

84. *Basra, Iraq, March 30, 2003:* A British soldier enjoys a cup of tea with an Iraqi man.
Dan Chung/AP Wide World Photos

85. *Baghdad, Iraq, April 4, 2003:* An Iraqi child plays with a toy gun in the Shiite shrine.
Jerome Delay/AP Wide World Photos

86–187. *Baghdad, aired April 4, 2003:* Television footage of Saddam Hussein touring the al-Mansour residential neighborhood of the capital.
Iraqi Television/Agence France Presse

88–189. *Baghdad, April 4, 2003:* U.S. Army soldiers enter the main terminal of Baghdad International Airport.
Scott Nelson/Getty Images

90–191. *Kuwait, April 5, 2003:* Senior Aircraftswoman (SAC) Caroline Richards (top left), SAC chef Jo Gemmell (top right), Corporal Katherine McNaught (center), Corporal Natalie Dornan (bottom left), and SAC Jo Davies pose in their uniforms and in their civilian clothes.
Russell Boyce/Agence France Presse

92–193. *Outside of Basra, Iraq, April 5, 2003:* British AS 90 155mm self-propelled guns fire on Iraqi targets.
Dan Chung/Reuters

94. *South of Baghdad, April 5, 2003:* U.S. Army soldiers pose in front of a mosaic of Saddam Hussein outside the Iraqi Republican Guard Medina Division headquarters.
John Moore/AP Wide World Photos

95. *South of Baghdad, April 5, 2003:* A US Army armored vehicle smashes a mosaic of Saddam Hussein outside the Iraqi Republican Guard Medina Division headquarters.
John Moore/AP Wide World Photos

96–197. *Baghdad, April 5, 2003:* The Iraqi minister of information Mohammed Sahid Al Saha addresses the media at a news conference.
Samir Mezban/AP Wide World Photos

98–199. *Baghdad, April 5, 2003:* Two injured Iraqi boys await treatment at Al Kindi hospital.
Bruno Stevens/Cosmos/Aurora

00–201. *Baghdad, April 6, 2003:* Iraqi soldiers celebrate on a destroyed U.S. Abrams tank in the southern outskirts of the capital.
Bruno Stevens/Cosmos/Aurora

02–203. *Baghdad, April 11, 2003:* A severely wounded Iraqi boy whose family was killed during a coalition air strike.
Jobard/Sipa

04–205. *Baghdad, April 4, 2003:* Ali Kiam, a surgeon at Yarmouk Hospital, takes a cigarette break between operations.
Jerome Delay/AP Wide World Photos

06–207. *Outskirts of Baghdad, April 6, 2003:* U.S. Marines exchange fire during a battle for a key bridge into the capital.
Laurent Rebours/AP Wide World Photos

08–209. *Somewhere over Iraq, April 4, 2003:* An F-14A Tomcat is refueled by an Air Force KC-135 tanker.
U.S. Navy/AP Wide World Photos

10–211. *Baghdad, April 6, 2003:* Iraqi medical staff work by candlelight as power outages darken much of the capital.
Faleh Kheiber/Reuters

12–213. *Baghdad, April 6, 2003:* Two U.S. Marines take cover in an alcove of a building as Iraqi mortar fire explodes nearby.
Timothy Fadek/Polaris

14–215. *Outside of Baghdad, April 6, 2003:* Sergeant Paul Ingram (left) is greeted by Sergeant Thomas Slago after a firefight with Iraqi irregular forces south of the capital.
John Moore/AP Wide World Photos

216–217. *Baghdad, March 24, 2003:* An Iraqi man, whose mother was killed by an allied air strike, waits outside the morgue for her body to be released for the funeral.
Jerome Delay/AP Wide World Photos

218–219. *Tikrit, Iraq, April 14, 2003:* A U.S. Marine arrests an Iraqi soldier.
Marco Di Lauro/Getty Images

220–221. *Arabian Gulf, April 7, 2003:* A U.S. support helicopter transfers equipment from the *USS Abraham Lincoln* to the *USS Nimitz.*
Yesenia Rosas/U.S. Navy/AP Wide World Photos

222–223. *Baghdad, March 28, 2003:* Members of the Amer family embrace near the remains of family members who were killed during an allied air strike.
Jerome Delay/AP Wide World Photos

224. *Outskirts of Baghdad, April 7, 2003:* Corporal Christopher Tate awaits orders on the southeastern edge of the capital.
Laura Rauch/AP Wide World Photos

225. *Basra, Iraq, April 7, 2003:* An Iraqi civilian walks past fallen signs.
Pauline Lubens/KRT

226–227. *Basra, Iraq, April 7, 2003:* An Iraqi man yells during looting at Basra University.
David Butow/Corbis SABA

228–299. *Outskirts of Baghdad, April 7, 2003:* U.S. Marines rush to secure the damaged Baghdad Highway Bridge.
Kuni Takahashi/*Boston Herald*/ReflexNews

230–231. *USS* Kitty Hawk, *Persian Gulf, April 4, 2003:* An F-14A Tomcat's afterburners light up the evening sky as it is launched from the deck of the USS *Kitty Hawk.*
Steve Helber/AP Wide World Photos

232–233. *Baghdad, April 7, 2003:* U.S. Army soldiers secure one of Saddam Hussein's palaces.
John Moore/AP Wide World Photos

234–235. *Baghdad, April 11, 2003:* U.S. Army soldiers and armored vehicles secure Saddam Hussein's military parade grounds.
John Moore/AP Wide World Photos

SADDAM HUSSEIN'S ART & PHOTO ALBUM

238. (top) *Location and date unknown:* Saddam Hussein playing with his youngest daughter, Hala.
Abaca Press/KRT

238. (bottom) *Location and date unknown:* Saddam Hussein on a family outing.
Abaca Press/KRT

239. (top) *Location unknown, 1990:* Saddam Hussein, center, seated with family members: (back row, from left) Hussein Kamel; Saddam Kamel; his wife and Saddam Hussein's daughter Rana; Saddam's eldest son, Uday; Hussein Kamel's wife and Saddam Hussein's eldest daughter, Raghda, with her son Ali, Saddam Hussein's first grandson.
AP Wide World Photos

239. (bottom) *Baghdad, April 16, 2003:* A fantasy painting from the wall of one of Saddam Hussein's safe houses.
John Moore/AP Wide World Photos

240. *Location and date unknown:* In this photo, which was found in one of Uday's homes on April 14, 2003, Saddam Hussein pets a tiger.
AP Wide World Photos

241. (top) Saddam Hussein.
AP Wide World Photos

241. (middle) Saddam's eldest son, Uday.
AP Wide World Photos

241. (bottom) Saddam's second son, Qusay.
AP Wide World Photos

242. *Date unknown:* A portrait of Saddam Hussein and his wife found in one of the Iraqi president's palaces.
Patrick Baz/Agence France Presse

43. *Baghdad, April 16, 2003:* A fantasy painting from the wall of one of Saddam Hussein's safe houses.
John Moore/AP Wide World Photos
44. (top) *Baghdad, date unknown:* Saddam's half brother, on crutches, and Saddam's son Uday in Saddam International Airport.
Cosmos/Aurora
44. (bottom) *Baghdad, December 3, 2002:* An Iraqi presidential guard stands in front of Al-Sajoud palace.
Jerome Delay/AP Wide World Photos
45. (top) *Baghdad, April 7, 2003:* Interior of one of Saddam Hussein's palaces.
AP Wide World Photos/APTN
45. (bottom) *Baghdad, date unknown:* Saddam Hussein kisses an unidentified woman.
Cosmos/Aurora
46–247. *Baghdad, December 3, 2002:* Al-Sajoud palace.
Jerome Delay/AP Wide World Photos
48. *Baghdad, October 30, 1997:* Saddam Hussein inspects weapons presented to him as gifts.
AP Wide World Photos/Iraq News Agency
49. (top) *Baghdad, date unknown:* Uday's wedding.
Cosmos/Aurora
49. (bottom) *Baghdad, December 3, 2002:* Interior of one of Saddam Hussein's palaces.
Jerome Delay/AP Wide World Photos
50. (top) *Baghdad, date unknown:* Saddam Hussein (center) with his three half brothers and son Uday (far right).
Cosmos/Aurora
50. (bottom) *Baghdad, December 3, 2002:* Journalists visit the foyer of Al-Sajoud palace.
Jerome Delay/AP Wide World Photos
51. (top) *Location and date unknown:* Saddam Hussein kisses an unidentified woman.
Cosmos/Aurora
51. (bottom) *Baghdad, date unknown:* Saddam Hussein (center) with his three half brothers and Uday's bride.
Cosmos/Aurora
52–253. *Baghdad, April 16, 2003:* Fantasy paintings from the walls of one of Saddam Hussein's safe houses.
John Moore/AP Wide World Photos
54. (top) *Baghdad, April 28, 1997:* Saddam Hussein cuts the cake as part of the celebrations for his sixtieth birthday.
AP Wide World Photos/Iraqi News Agency
54. (bottom) *Babylon, Iraq, March 23, 1998:* A palace of Saddam Hussein sits on a hilltop above the ruins of ancient Babylon.
Jockel Finck/AP Wide World Photos
55. *Baghdad, December 12, 2000:* Saddam Hussein's eldest son, Uday, hunts for wild birds on the Tigris River.
AP Wide World Photos
56–257. *Baghdad, April 16, 2003:* Fantasy paintings from the walls of one of Saddam Hussein's safe houses.
John Moore/AP Wide World Photos

HE FALL OF BAGHDAD

60–261. *Outside Baghdad, April 7, 2003:* Iraqis celebrate south of Baghdad.
Michael Macor/*San Francisco Chronicle*/Corbis SABA
62–263. *Baghdad, April 8, 2003:* Saddam mural with graffiti from U.S. troops.
Michael Macor/*San Francisco Chronicle*/Corbis SABA
64–265. *Medina, Iraq, April 8, 2003:* Local children on bikes race British tanks as allied forces secure the town.
Bruce Adams/AP Wide World Photos
266. *Basra, Iraq, April 8, 2003:* Iraqis beat a suspected Fedayeen.
Jon Mills/Agence France Presse
267. *Baghdad, April 8, 2003:* Fedayeen flee with their wounded.
Patrick Robert/Corbis
268–269. *Basra, Iraq, April 7, 2003:* Iraqi women celebrate on the northern outskirts of Basra.
Chis Ison/Reuters
270–271. *Basra, Iraq, April 8, 2003:* An Iraqi girl cries as British tank forces move in on the Baath party office.
Odd Andersen/Agence France Presse
272–273. *Baghdad, April 12, 2003:* A young Iraqi on a swing set in a damaged neighborhood.
Heidi Levine/SIPA
274–275. *Baghdad, April 9, 2003:* U.S. Army Sergeant John Beck is welcomed by Iraqi civilians.
John Moore/AP Wide World Photos
276–277. *Baghdad, April 7, 2003:* U.S. Army soldiers search one of Saddam Hussein's palaces, which was damaged by allied air strikes.
John Moore/AP Wide World Photos
278–279. *Basra, Iraq, April 7, 2003:* Iraqi looters scale a fence.
Paolo Pellegrin/Magnum Photos
280–281. *Iraq, April 16, 2003:* General Tommy Franks walks through the remains of one of Saddam Hussein's palaces.
Corbis
282–283. *Baghdad, April 7, 2003:* U.S. Army Staff Sergeant Chad Touchett (center) relaxes with comrades following a search of one of Saddam Hussein's palaces.
John Moore/AP Wide World Photos
284. *Basra, Iraq, April 8, 2003:* Iraqi children stomp on a portrait of Saddam Hussein that was taken down from the town center.
Tony Nicoletti/AP Wide World Photos
285. *Fayetteville, North Carolina, April 11, 2003:* Taylyn Solomon at the funeral of his father, Sergeant Roderic A. Solomon, who was killed near the town of Najaf, Iraq.
Ellen Ozier/Reuters
286–287. *Al Hilla, Iraq, April 1, 2003:* Karem Mohammed weeps over the bodies of his family killed in an allied airstrike.
Ali Heider/AP Wide World Photos
288–289. *Baghdad, April 19, 2003:* Unfinished statues of Saddam Hussein and others in the studio of one of Hussein's personal sculptors.
Timothy Fadek/Polaris
290–291. *Mossul, Iraq, April 13, 2003:* Iraqi children enjoy a swim in the private lake in front of one of Saddam Hussein's palaces.
Patrick Barth/Getty Images
292–293. *Baghdad, April 9, 2003:* U.S. Marines advance on the headquarters of the Fedayeen.
Laura Rauch/AP Wide World Photos
294–295. *Baghdad, April 13, 2003:* Iraqis throw stones at a chandelier as they loot one of Saddam Hussein's palaces.
Heidi Levine/SIPA
296–297. *Baghdad, April 9, 2003:* A portrait of Saddam Hussein hangs on the burning Ministry of Transport and Communication building.
Jerome Delay/AP Wide World Photos
298–299. *Baghdad, April 11, 2003:* Iraqi women mourn.
Jobard/Sipa
300–301. *Baghdad, April 9, 2003:* U.S. Marines take position at the Martyrs Monument.
Oleg Popov/Reuters
302. *Baghdad, April 9, 2003:* A statue of Saddam Hussein stands in front of the burning offices of his son Uday.
Daily Mirror Gulf coverage/Mirrorpix

303. (top) *Baghdad, April 9, 2003:* U.S. Marines cover the face of a Saddam Hussein statue with an American flag.
Daily Mirror Gulf coverage/Mirrorpix
303. (middle) *Baghdad, April 9, 2003:* U.S. Marines prepare to pull down a statue of Saddam Hussein.
Ramzi Haidar/Agence France Presse
303. (bottom) *Baghdad, April 9, 2003:* U.S. Marines cover the face of a Saddam Hussein statue with a pre-1991 Iraqi flag.
Patrick Baz/Agence France Presse
304–305. (top) *Baghdad, April 9, 2003:* A forty-foot statue of Saddam Hussein is pulled down in Ferdowsi square.
APTN/AP Wide World Photos
304–305. (bottom) *Kerbala, Iraq, April 6, 2003:* Iraqi citizens and U.S. soldiers pull down a statue of Saddam Hussein.
Peter Andrews/Reuters
306–307. *Baghdad, April 9, 2003:* Iraqis destroy a statue of Saddam Hussein after it was pulled down by a U.S. armored vehicle.
Karim Sahib/Agence France Presse
308–309. *Baghdad, April 9, 2003:* Iraqis loot furniture while walking past a torn portrait of Saddam Hussein.
John Moore/AP Wide World Photos
310–311. *Basra, Iraq, April 9, 2003:* An Iraqi boy helps himself to sugar from a warehouse.
Jon Mills/AP Wide World Photos
312–313. *Outside of Basra, Iraq, April 9, 2003:* British troops cool down under an improvised shower.
Giles Penfold/Reuters
314–315. *Baghdad, April 10, 2003:* Iraqi women peek from their doorway at a U.S. Marine taking a break from patrolling.
Kuni Takahashi/*Boston Herald*/ReflexNews
316–317. *Basra, Iraq, April 10, 2003:* British officers meet with local Imams.
Mark Richards/*Daily Mail*/AP Wide World Photos
318–319. *Baghdad, April 8, 2003:* A1C Nick Taylor (left) and Specialist Bryan Slick play Ping-Pong during a day off at one of Saddam Hussein's palaces.
Sobhani/*San Antonio Express*/Zuma Press
320. *Baghdad, April 12, 2003:* U.S. Army armored vehicles pass a group of looters.
John Moore/AP Wide World Photos
321. *Mosul, Iraq, April 11, 2003:* An Iraqi boy collects money in front of a bank.
Joseph Barrak/Agence France Presse
322–323. *Baghdad, April 13, 2003:* Iraqi National Museum director Mushin Hasan amid the ruins of his museum, which was severely looted.
Mario Tama/Getty Images
324–325. *Baghdad: April 15, 2003:* The remains of Saddam Hussein's vintage car collection, which had been looted.
Hyoung Chang/*Denver Post*/Polaris
326–327. *Baghdad, April 12, 2003:* An Iraqi boy picks up bullet casings.
Olivier Coret/*In Visu*/Corbis
328–329. *Baghdad, April 10, 2003:* An Iraqi looter displays money notes taken from a downtown bank.
Laurent Rebours/AP Wide World Photos
330–331. *Baghdad, April 13, 2003:* Imam Hassin Abraham and elder Muslims speak with reporters in front of the Hassin Mosque.
Kuni Takahashi/*Boston Herald*/ReflexNews
332–333. *Baghdad, April 8, 2003:* PFC Mark Bennett fishes in a lake surrounding one of Saddam Hussein's palaces.
B. M. Sobhani/*San Antonio Express*/Zuma Press
334–335. *Baghdad, April 13, 2003:* An Iraqi looter takes stacks of china from the bomb blasted dome of Saddam Hussein's Al-Salam presidential palace.
David Guttenfelder/AP Wide World Photos
336–337. *Baghdad, April 14, 2003:* A painting taken from the tales of the *Thousand and One Nights* hangs in Uday's wing of Saddam Hussein's main presidential palace.
Ramzi Haidar/Agence France Presse
338–339. *Basra, Iraq, April 7, 2003:* Royal Marine Stuart Lawley stands in a corridor of one of Saddam Hussein's palaces.
PA Photos
340. *Baghdad, April 13, 2003:* U.S. Army Specialist David Buchanan poses in a huge chair in one of Saddam Hussein's palaces.
Romeo Gacad/Agence France Presse
341. Iraq's Most Wanted playing cards.
Command Central/Dept. of Defense
342–343. *Baghdad, April 12, 2003:* An Iraqi looter attempts to pull down a chandelier inside one of Saddam Hussein's palaces.
Kuni Takahashi/*Boston Herald*/ReflexNews
344–345. *Baghdad, April 10, 2003:* U.S. infantrywoman Felicia Harris on a bed in Uday's palace.
Mike Moore/Mirrorpix
346–347. *Baghdad, April 12, 2003:* An Iraqi looter takes a couch from an upper-class home.
Benjamin Lowy/Corbis
348–349. *Baghdad, April 16, 2003:* Iraqi looters take a break outside of Saddam Hussein's main presidential palace.
Christophe Calais/*In Visu*/Corbis
350–351. *Umm Qasr, Iraq, April 12, 2003:* A British soldier and an Iraqi girl speak in a classroom.
Graeme Robertson/Getty Images
352–353. *Baghdad, April 10, 2003:* U.S. infantrymen at the pool bar of Uday's palace.
Daily Mirror/Mirrorpix
354–355. *Baghdad, March 24, 2003:* Photos of U.S. POWs taken from Iraqi television after their convoy was ambushed near Nasiriyah, Iraq. They were later rescued.
AP Wide World Photos/Iraqi TV via APTN
356–357. *El Paso, Texas, April 19, 2003:* Former POWs Joseph Hudson (left) and Patrick Miller celebrate their return to America after landing at Fort Bliss.
Ellis Neel/Corbis
358–359. *Baghdad, April 12, 2003:* U.S. soldiers attend a memorial ceremony for six soldiers killed during combat operations.
Handout/Getty Images
360–361. *Baghdad, April 9, 2003:* Iraqis celebrate with a U.S. Marine during the fall of Baghdad.
Timothy Fadek/Polaris
362–363. *Baghdad, April 12, 2003:* An Iraqi man offers a carnation to U.S. Marine Sergeant Jose Acostavelo as his squad patrols a neighborhood in the capital.
Julie Jacobson/AP Wide World Photos
364–365. *Kirkuk, Northern Iraq, April 15, 2003:* An Iraqi man passes by a painted wall.
Patrick Barth/Getty Images
366–367. *Baghdad, April 14, 2003:* U.S. Army helicopters patrol the skies above Baghdad.
Jobard/SIPA